AF553777

101 GREAT WOMEN WHO SHAPED THE WORLD

101 GREAT WOMEN WHO SHAPED THE WORLD

KAMAL S. SRIVASTAVA
SANGEETA SRIVASTAVA

A P H PUBLISHING CORPORATION
4435-36/7, ANSARI ROAD, DARYA GANJ
NEW DELHI-110 002

Published by
S.B. Nangia
A.P.H. Publishing Corporation
4435–36/7, Ansari Road, Darya Ganj,
New Delhi-110002
Phone: 011–23274050
e-mail: aphbooks@gmail.com

2026

Typeset by
Ideal Publishing Solutions
C-90, J.D. Cambridge School,
West Vinod Nagar, Delhi-110092

Printed at
DIVINE DIGITAL PRINTERS
Ansari road Daryaganj Delhi-110002

Contents

Preface

A great woman or personality or celebrity is a female person who has a prominent profile in history due to certain reasons and is frequent talked about in the media. This single volume publication provides readers with the detailed biographies of important women throughout history, worldwide. A list of 101 prominent female individuals from all walks of life, who have made significant contributions and impact on world history and development, have been selected for each individual entry.

This book titled "101 Great Women Who Shaped the World" provides readers with the details of most influential women who shaped the world. With the help of this publication one can learn all about the many famous and fascinating women in human history. The publication includes biography of following: Audrey Hepburn; Indira Gandhi; Queen Victoria; Madonna; Mary Magdalene; Benazir Bhutto; Jacqueline Kennedy Onassis; Cleopatra V of Egypt; Marilyn Monroe; Mother Teresa; Anne Frank; Sappho; Oprah Winfrey; Billie Jean King; Eleanor of Aquitaine; Hillary Rodham Clinton; Brigitte Bardot; Eleanor Roosevelt; Teresa of Ávila; Agatha Christie; Catherine de' Medici; Mata Hari; J.K. Rowling; Queen Elizabeth I; Ingrid Bergman; Catherine The Great; Mary Wollstonecraft; Jane Austen; Aung San Suu Kyi; Harriet Beecher Stowe; Mae West; Florence Nightingale; Susan B. Anthony; Emily Dickinson; Marie Antoinette; Emily Pankhurst; Marie Curie; Emily Murphy; Luxemburg, Rosa; Zsa Zsa Gabor; Bette Davis; Helena Rubinstein; Helen Keller; Coco Chanel; Emily Bronte; Amelia Earhart; Raisa Gorbachova; Katharine Hepburn; Simone de Beauvoir; Marlene Dietrich; Rosa Parks; Jiang Qing; Hildegard of Bingen; Billie Holiday; Eva Peron; Yoko Ono; Betty Friedan; Queen Elizabeth II; Margaret Thatcher; Dr. Dian Fossey; Germaine Greer; Enid Blyton;

Betty Williams; Lauren Bacall; Julie Andrews; Jane Goodall; Amy Johnson; Martina Navratilova; Daphne du Maurier; Golda Meir; Tegla Loroupe; Elizabeth Taylor; Virginia Woolf; Maya Angelou; Anita Roddick; Dame Barbara Cartland; Beatrix Potter; Shakira; Anna Pavlova; Lakshmibai; Sonia Gandhi; Edith Cavell; St Therese of Lisieux; Estée Lauder (Person); Marie Stopes; Susan Boyle; Iris Murdoch; Vera Brittain; Whoopi Goldberg; Sarojini Naidu; Gabriela Mistral; Catherine 'Kate' Middleton; Mary Seacole; Kasturba Gandhi; Maria Callas; Mirabai; Lady Gaga; Pratibha Patil; Boudica; Paula Radcliffe and Louisa May Alcott.

- Editor

1

Audrey Hepburn

Early Life

Audrey Hepburn was born to an English father and Dutch mother in Belgium, May 4th, 1929. Her father's job as an insurance agent meant the family often moved between England, Holland and Belgium. In 1935, her parents divorced; one reason for this was that her father was a Nazi sympathiser. The divorce was very traumatic for six year old Audrey; she would later say it was the most traumatic incident of her life. After the war, despite suffering under the Nazi occupation, Audrey later tracked down her father to Dublin and supported him financially. From 1935-58, Audrey went to boarding school in Kent; in 1939 her mother moved the family to Arnhem in the Netherlands, where she thought it would be safe from Nazi invasion. However, in 1940, the Netherlands was overrun and the country fell under Nazi occupation until liberation in 1945. During this time, Audrey went to school at the Arnhem conservatory where she studied and also learnt ballet. At one time she considered taking ballet as a serious career occupation. During the occupation, it was said she would often dance in various locations, helping to raise money for the underground movement.

Towards the end of the war, the occupation of Netherlands became increasingly brutal. After the D Day landings of 1944, the Germans took most of the pitiable rations of the Dutch, leaving many to starve or freeze to death. Reprisals and shootings against the local population were common. As a young girl, Audrey saw her uncle and mother's cousin shot in the street by the Germans. She also recalls seeing a train load of Jewish children being herded into cattle trucks for deportation.

> "I have memories. More than once I was at the station seeing trainloads of Jews being transported, seeing all these faces over the top of the wagon."

The harrowing experiences of war left a profound mark on Audrey; it was one reason for her later commitments to the UNICEF children's charity.

> *"I can testify to what UNICEF means to children, because I was among those who received food and medical relief right after World War II"*

She felt a natural empathy and sympathy for children who were the victims of war and starvation. During the war Audrey suffered anaemia, respiratory problems and edema (swelling of the limbs) Audrey later noted a similarity between her wartime experience and that of Anne Frank. She read her diary in 1946, and said it "left her feeling devastated." However, despite the ongoing horrors of the occupation, Audrey passed her time through drawing and practising ballet. After the war, Audrey went to London where she continued to practise ballet. She had great talent but her height and malnutrition during the war meant that she was unable to become a really great ballerina, and so decided to seek work as an actor.

Audrey Hepburn Acting

After several minor roles in films such as *The Lavender Hill Mob*, Audrey was chosen to play *Gigi* a hit West end play. She was given an award by the Theatre world for best debut performance. Her first film was *Secret People* in 1952; a film about a prodigy ballerina, which was a natural choice for Audrey to play. It was during the filming for this that she was spotted by director William Wyler. He was producing a film "*Roman Holiday*" and he felt the innocence and elfin beauty of Audrey Hepburn would make a perfect choice for the English Princess, who spends a day in Rome in the company of Gregory Peck. The film was a great hit and on the advertising Audrey Hepburn was given the same billing as Gregory Peck. In many respects, Audrey outshone her more illustrious lead Gregory Peck; as Gregory Peck predicted it was Audrey who would be given an Oscar for her performance. This film established her place in Hollywood's elite and allowed her to play against many of the leading men of the time. For example, *Sabrina* with Humphrey Bogart and Fred Astaire in Funny Face.

Enduring Appeal of Audrey Hepburn

The enduring popularity and appeal of Audrey Hepburn can be attributed to many factors. She had a natural beauty and elegance; she has often been voted the most beautiful woman of the century. However she also had an aura of childlike innocence which portrayed a natural charm and humour. She was held in high regard by many in the film industry; she avoided conflict and many of the top actors said how much they enjoyed working with Audrey.

Although she was one of Hollywood's great stars of the 1950s and 1960, she didn't allow her fame to go to her head; often she would be quite happy to stay at home with her family. Her son wrote a moving tribute to his mother in this books *"Audrey Hepburn, an Elegant Spirit: A son remembers"* In 1961, Audrey played one of her most demanding roles, the extrovert Holly Golightly in *"Breakfast at Tiffany's"* She said of her role that it was "one of the jazziest of my career" she said it was contrary to her introverted nature and thus was quite difficult to pull off. However, her performance was one of the most iconic roles of the 1960s. The film has retained an enduring popularity to this day. In 2006, the "little black dress" from the film was sold in auction for just under £0.5 million. The proceeds were given to one of Audrey's charities. In 1959, she stared in *"The Nun's Story"* - quite different to her other roles, this was a challenging portrayal of a young nun, Sister Luke, who trained to be a novice nun before spending time as a missionary in the Congo. Sister Luke also faces a painful spiritual dilemma as she returns to Belgium and the Nazi occupation. With some parallels to her own life, the film showed the multi faceted acting talents of Audrey Hepburn.

Audrey Hepburn UNICEF Charity Work

From 1967, after 15 years in film, she acted only occasionally. She spent more time with her family and also working with UNICEF. She was appointed as a special ambassador to UNICEF and became actively involved in campaigns to improve conditions for children around the world. In 1988 she visited Ethiopia at a camp for children on seeing the poverty and starvation she remarked:

> "I have a broken heart. I feel desperate. I can't stand the idea that two million people are in imminent danger of starving to death, many of them children, [and] [sic] not because there isn't tons of food sitting in the northern port of Shoa."
>
> —*Audrey Hepburn*

She also visited street children in South America and was appalled to see children living in such conditions. She later reported to Congress how UNICEF had been able to make a difference

> "I saw tiny mountain communities, slums, and shantytowns receive water systems for the first time by some miracle-and the miracle is UNICEF. "I watched boys build their own schoolhouse with bricks and cement provided by UNICEF."

Death of Audrey Hepburn

After returning from Somalia in 1992 Audrey Hepburn developed cancer of the colon. The disease proved to be untreatable in January 1993 she died in Switzerland aged 63.

2

Indira Gandhi

Indira Priyadarshini Gandhi (19 November 1917 - 31 October 1984) was the Prime Minister of the Republic of India for three consecutive terms from 1966 to 1977 and for a fourth term from 1980 until her assassination in 1984, a total of fifteen years. She was India's first and, to date, only female Prime Minister. In 1999, she was voted the greatest woman of the past 1000 years in a poll carried by BBC news, ahead of other notable women such as Queen Elizabeth I of England, Marie Curie and Mother Teresa. Born in the politically influential Nehru dynasty, she grew up in an intensely political atmosphere. Despite the same last name, she was of no relation to the statesman Mohandas Gandhi. Her grandfather, Motilal Nehru, was a prominent Indian nationalist leader. Her father, Jawaharlal Nehru, was a pivotal figure in the Indian independence movement and the first Prime Minister of Independent India. On returning to India from Oxford in 1941, she became involved in the Indian Independence movement. In the 1950s, she served her father unofficially as a personal assistant during his tenure as the first Prime Minister of India. After her father's death in 1964, she was appointed as a member of the Rajya Sabha by the President of India and became a member of Lal Bahadur Shastri's cabinet as Minister of Information and Broadcasting.

The then Congress Party President, K. Kamaraj was instrumental in making Indira Gandhi the Prime Minister after the sudden demise of Shastri. Gandhi soon showed an ability to win elections and outmaneuver opponents through populism. She introduced more left-wing economic policies and promoted agricultural productivity. A decisive victory in the 1971 war with Pakistan was followed by a period of instability that led her to impose a state of emergency in 1975; she paid for the authoritarian excesses of the period with three years in opposition. Returned to office in 1980, she became increasingly involved in an escalating conflict with separatists in Punjab that eventually led to her assassination by her own bodyguards in 1984.

Legacy

Being the first woman Prime Minister of India, and an influential leader; in a prevalently male-dominated society, Indira Gandhi is a symbol of feminism in India.. As per economic surveys, when Indira Gandhi became Prime Minister, 65% of the country's population was below the poverty line, and when her regime ended in 1984, this figure was 45%. During her rule, food production increased by 250%. Literacy also increased in India by 30%.

The goodwill of the rural population earned by Gandhi still has its effects on the success of the Congress Party in rural India, as well as the popular support of the Nehru-Gandhi Family. She is reverently remembered in many parts of rural India as Indira-*Amma* ("*Amma*" means "mother" in many Indian languages). Her *Garibi Hatao* slogan is still used by the Congress during political campaigns. The present president of the Indian National Congress, Sonia Gandhi, who is also the daughter-in-law of Indira Gandhi, is said to style herself in resemblance to her.

The Indira Awaas Yojana, a programme of the central government to provide low-cost housing to rural poor, is named after her. The international airport at New Delhi is named the Indira Gandhi International Airport in her honour.

A negative legacy Mrs. Gandhi will be associated with is that of fostering a culture of nepotism.

3

Queen Victoria

Queen Victoria was born 24 May, 1819. She was the granddaughter of George III, and her father, Edward was fourth in line to the throne. But when the prince of Wales died early, his brothers sought to get married and maintain the line of succession. Edward married Princess Victoria from Germany and the couple had just one child, Alexandrina Victoria, who was born at Kensington Palace in 1819. As a young girl, Victoria's father died, followed 6 days later by King George III. The throne then passed to King William IV, but, he too died early. This left Victoria to be crowned at the age of 18, in June 1837. Queen Victoria was to reign until her death on 22nd January 1901.

Queen Victoria and Nineteenth Century Britain

The 19th Century was a time of unprecedented expansion for Britain in term of both of industry and Empire. Although her popularity ebbed and flowed during her reign, towards the end of her crown, she had become a symbol of British imperialism and pride. The Victorian period also witnessed great advances in science and technology. It became known as the steam age, enabling people to easily travel throughout the UK and the World. Queen Victoria was emblematic of this period. She was an enthusiastic supporter of the British Empire. She celebrated at Lord Kitchener's victory in the Sudan, she supported British involvement in the Boer War. She was also happy to preside over the expansion of the British Empire, which was to stretch across the globe. In 1877 Queen Victoria was made Empress of India, in a move instigated by the imperialist Disraeli. Famously, at the end of the Victorian period, people could say 'the sun never set on the British Empire'. Queen Victoria was conservative in her politics and social views. This led to an unfortunate episode. When she saw a servant who appeared to be pregnant, Victoria claimed she was having an affair. The Queen actually made her take a test to prove she was a virgin. The test was positive and the growth in her stomach was actually a form of cancer; a few months later the servant died and Queen Victoria suffered a decline in her popularity as a result of this episode.

In the early part of her reign she become a close friend and confident of the Prime Minister, Lord Melbourne. She spent many hours talking to him and relied on his political advice. Lord Melbourne was a whig, with conservative attitudes. He tried to shield Queen Victoria from the extreme poverty that was endemic in parts of the UK. Queen Victoria was also highly devoted to her husband, Prince Albert; together they had 9 children. When Prince Albert died in 1861, at the age of 41, Queen Victoria went into deep mourning and struggled to overcome this loss. She became reclusive and was reluctant to appear in public. Parliament and Benjamin Disraeli had to use all their persuasive power to get her to open parliament in 1866 and 1867. This hiding from the public led to a decline in popularity. However, by the end of her reign, her popularity was restored. This was partly due to the rise of Great Britain as the leading super power of the era. For various reasons, several attempts were made on the life of Queen Victoria. These were mostly between 1840 and 1882. She was always unharmed, but, her courageous attitude helped to endear her to the public.

4

Madonna

Madonna, born Louis Cicconi, in 1958, was brought up in Rochester Hills Michigan. In her early life she worked as a waitress and went to New York to learn modern dance. In 1982 she released her first single *'Everybody'* and in 1983, she launched her first album *'Madonna'* It sold very well, but it was her next album *'Like a Virgin'* which made her into an international superstar. The album sold over 12 million copies, helped by the hit single *'Like a Virgin'* which stayed at number one for 6 weeks. As well as being a successful music recording artist, Madonna was very influential in influencing fashion and attitudes to life. Her trademark looks included fish net stockings, a Christian cross, bleached hair and Capri skirts. Throughout her career, Madonna has courted controversy for her mix of sexuality and religious imagery. Her attitudes sparked criticism from the Vatican and the Vatican discouraged people from attending her concerts because of the eroticism. Madonna, generally, remained unapologetic and continued to perform her routines.

After forming her own company - Madonna, produced a book titled *'Sex'*. It featured nude photography and was quite controversial, especially in US, but sold over half a million copies. She has acted in a couple of films, without much commercial success though her appearance in Evita about Eva Peron, the famous Argentinean wife of the President received good reviews. She has now produced over 10 albums and remains a very influential artist still capable of hitting the top of the album charts. Her early Catholic roots influenced her music and videos, but, in the early 1990s she embraced Kabbalah a Jewish mystical sect.

5

Mary Magdalene

Mary Magdalene (flourished 1st century AD) was one of Jesus' most celebrated disciples. She is famous because she is said to have been the first person to see Jesus after he rose from the dead, according to John 20 and Mark 16:9, part of the so-called "Longer Ending" to that chapter. According to Luke 8:2 and Mark 16:9, Jesus cleansed her of "seven demons." Some contemporary scholars contend this concept means *healing from illness*, not forgiveness of sin. On the other hand, some major Christian saints, including St. Bede and St. Gregory, interpret the seven devils to signify that she was "full of all vices." Hence, on this interpretation, the episode does signify the forgiveness of sins. Mary Magdalene is the leader of a group of female disciples believed to have been present at the cross after the male disciples (excepting John the Beloved) had fled, and at his burial. Mary was a devoted follower of Jesus, entering into the close circle of those taught by Jesus during his Galilean ministry. She became prominent during the last days, accompanying Jesus during his travels and following him to the end. She witnessed his Crucifixion and burial.

Mary Magdalene is referred to in early Christian writings as "the apostle to the apostles." In apocryphal texts, she is portrayed as a visionary and leader of the early movement, who was loved by Jesus more than the other disciples. Several Gnostic gospels, such as the Gospel of Mary, written in the early 2nd century, see Mary as the special disciple of Jesus who has a deeper understanding of his teachings and is asked to impart this to the other disciples. Mary Magdalene is considered by the Roman Catholic, Eastern Orthodox, Anglican, and Lutheran churches to be a saint, with a feast day of July 22. The Eastern Orthodox churches also commemorate her on the Sunday of the Myrrhbearers which is the second Sunday after Pascha (Easter). She is also an important figure in the Bahá'í Faith.

Pistis Sophia

Pistis Sophia, possibly dating as early as the 2nd century, is the best surviving of the Gnostic writings. *Pistis Sophia* presents a long dialog with Jesus in the form of his answers to questions from his disciples. Of the 64 questions, 39 are presented by a woman who is referred to as Mary or Mary Magdalene.

Gospel of Mary

Gospel of Mary is usually dated to about the same period as that of the Gospel of Philip. The Gospel was first discovered in 1896. The Gospel is missing six pages from the beginning and four in the middle. The identity of "Mary" appearing as the main character in the Gospel is sometimes disputed, but she is generally regarded to be Mary Magdalene. In the Gospel, Mary, presented here as one of the disciples, has seen a private vision from the resurrected Jesus and describes it to other disciples. Peter said to Mary, "Sister we know that the Savior loved you more than the rest of woman. Tell us the words of the Savior which you remember which you know, but we do not, nor have we heard them." Mary answered and said, "What is hidden from you I will proclaim to you." And she began to speak to them these words: "I, she said, I saw the Lord in a vision and I said to Him, Lord I saw you today in a vision."

Unfortunately, almost all of Mary's vision is within the lost pages. When Mary had said these things, she fell silent, since it was up to this point that the Savior had spoken to her. The repeated reference in the Gnostic texts of Mary as being loved by Jesus more than the others has been seen as supporting the theory that the Beloved Disciple in the canonical Gospel of John was originally Mary Magdalene, before a later redactor made changes in the Gospel.

Bahá'í Tradition

There are many references to Mary Magdalene in the sacred writings of the Bahá'í Faith. 'Abdu'l-Bahá, the son of the founder of the religion, said that she was "the channel of confirmation" to Jesus' disciples, a "heroine" who "re-established the faith of the apostles" and was "a light of nearness in his kingdom." 'Abdu'l-Bahá also wrote that "her reality is ever shining from the horizon of Christ," "her face is shining and beaming forth on the horizon of the universe forevermore" and that "her candle is, in the assemblage of the world, lighted till eternity." 'Abdu'l-Bahá considered her to be the supreme example of how women are completely equal with men in the sight of God and can at times even exceed men in holiness and greatness. Indeed he claimed that she surpassed all the men of her time, and that "crowns studded with the brilliant jewels of guidance" were upon her head. The Bahá'í writings also expand upon the scarce references to her life in the canonical Gospels, with a wide array of extra-canonical stories about her and sayings which are not recorded in any other extant historical sources. 'Abdu'l-Bahá claimed that Mary travelled to Rome and spoke before the Emperor Tiberius, which is presumably why Pilate was later recalled to Rome for his cruel treatment of the Jews (a tradition also attested to in the Eastern Orthodox Church). According to the memoirs of Juliet Thompson, 'Abdu'l-Bahá also compared Mary to Juliet, one of his most devoted followers, claiming that she even physically resembled her and that Mary Magdalene was Juliet Thompson's "correspondence in heaven."

Mary Magdalene, a Virgin after the Resurrection of Jesus Christ

Ambrose (*De virginitate* 3,14; 4,15) and John Chrysostom (*Matthew, Homily 88*) have suggested that Mary Magdalene was a virgin after the Resurrection of Jesus Christ.

Relationship with Jesus

The apocryphal Gospel of Philip depicts Mary as Jesus' *koinonos*, a Greek term indicating a "close friend" or "companion". Mary Magdalene is mentioned as one of three Marys "who always walked with the Lord" and as his companion

(Philip 59.6-11). The work also says that Lord loved her more than all the disciples, and used to kiss her often (63.34-36). The closeness described in these writings depicts Mary Magdalene, representing the Gnostics, as understanding Jesus and his teaching while the other disciples, representing the Church, did not. Kripal writes that "the historical sources are simply too contradictory and simultaneously too silent" to make absolute declarations regarding Jesus' sexuality. On the other hand, author John Dickson argues that it was common in early Christianity to kiss a fellow believer by way of greeting, and as such kissing would have no romantic connotations. Mary Magdalene appears with more frequency than other women in the canonical Gospels and is shown as being a close follower of Jesus. Mary's presence at the Crucifixion and Jesus' tomb, while hardly conclusive, is at least consistent with the role of grieving wife and widow.

Proponents of a married status of Jesus argue that it would have been unthinkable for an adult, unmarried Jew to travel about teaching as a rabbi. However, in Jesus' time the Jewish religion was very diverse and the role of the rabbi was not yet well defined. It was not until after the Roman destruction of the Second Temple in AD 70 that Rabbinic Judaism became dominant and the role of the rabbi made uniform in Jewish communities. Mary Magdalene is typically depicted in various ways in artwork.These depictions of her also show us how various artisits viewed her and Jesus' relationship. In Fra Angelico's painting noli me tangere, There is no sign of "suppressed desire" and there is not tension. Fra Angelico's painting (according to Robert Kiely) "is not a shocking or disturbing scene in which Mary overreacts, but a quiet beginning of a heavenly dance." Angolo Bronzino's painting 'noli me Tangere" is much more animated and "melodramatic." Mary and Jesus have much more emotion and there seems to be a tension and a scene being caused. In fact, people in the background are looking on to the scene as if it is scandalous. Both paintings have the title "Noli me Tangere" which means "do not touch me." It is said that after Jesus' resurrection, he said these words to Mary. Thus, a common debate is why he said these words to her.

6

Benazir Bhutto

Benazir Bhutto (21 June 1953 – 27 December 2007) was a Pakistani politician who chaired the Pakistan Peoples Party (PPP), a centre-left political party in Pakistan. Bhutto was the first woman elected to lead a Muslim state, having twice been Prime Minister of Pakistan (1988–1990; 1993–1996). She was Pakistan's first and to date only female prime minister. Her family is from the Bhutto tribe of Sindhis. Bhutto was the eldest child of former prime minister Zulfikar Ali Bhutto, a Pakistani of Sindhi descent and Shia Muslim by faith,Begum Nusrat Bhutto, a Pakistani of Iranian-Kurdish descent, similarly Shia Muslim by faith. Her paternal grandfather was Sir Shah Nawaz Bhutto, who came to Larkana District in Sindh before the independence from his native town of Bhatto Kalan, in the Indian state of Haryana. She studied Law at Lady Margaret Hall, Oxford University. After LMH, she studied at St Catherine's and became president of the Oxford Union in 1976

Bhutto was sworn in as Prime Minister for the first time in 1988 at the age of 35, but was removed from office 20 months later under the order of then-president Ghulam Ishaq Khan on grounds of alleged corruption. In 1993 she was re-elected but was again removed in 1996 on similar charges, this time by President Farooq Leghari. She went into self-imposed exile in Dubai in 1998.

Bhutto returned to Pakistan on 18 October 2007, after reaching an understanding with President Pervez Musharraf by which she was granted amnesty and all corruption charges were withdrawn.

On 27 December 2007, Bhutto was killed while leaving a campaign rally for the PPP at Liaquat National Bagh, where she had given a spirited address to party supporters in the run-up to the January 2008 parliamentary elections. After entering her bulletproof vehicle, Bhutto stood up through its sunroof to wave to the crowds. At this point, a gunman fired shots at her and subsequently explosives were detonated near the vehicle killing approximately 20 people. Bhutto was critically wounded and was rushed to Rawalpindi General Hospital. She was taken into surgery at 17:35 local time, and pronounced dead at 18:16.

Bhutto's body was flown to her hometown of Garhi Khuda Bakhsh in Larkana District, Sindh, and was buried next to her father in the family mausoleum at a ceremony attended by hundreds of thousands of mourners.

There was some disagreement about the exact cause of death. Bhutto's husband refused to permit an autopsy or post-mortem examination to be carried out. On 28 December 2007, the Interior Ministry of Pakistan stated that "Bhutto was killed when she tried to duck back into the vehicle, and the shock waves from the blast knocked her head into a lever attached to the sunroof, fracturing her skull". However, a hospital spokesman stated earlier that she had suffered shrapnel wounds to the head and that this was the cause of her death. Bhutto's aides have also disputed the Interior Ministry's account. On December 31, CNN posted the alleged emergency room admission report as a PDF file. The document appears to have been signed by all the admitting physicians and notes that no object was found inside the wound.

Al-Qaeda commander Mustafa Abu al-Yazid claimed responsibility for the attack, describing Bhutto as "the most precious American asset." The Pakistani government also stated that it had proof that al-Qaeda was behind the assassination. A report for CNN stated: "the Interior Ministry also earlier told Pakistan's Geo TV that the suicide bomber belonged to Lashkar i Jhangvi—an al-Qaeda-linked militant group that the government has blamed for hundreds of killings". The government of Pakistan claimed Baitullah Mehsud was the mastermind behind the assassination. Lashkar i Jhangvi, a Wahabi Muslim extremist organization affiliated with al-Qaeda that also attempted in 1999 to assassinate former Prime Minister Nawaz Sharif, is alleged to have been responsible for the killing of the 54-year-old Bhutto along with approximately 20 bystanders, however this is vigorously disputed by the Bhutto family, by the PPP that Bhutto had headed and by Baitullah Mehsud. On 3 January 2008, President Musharraf officially denied participating in the assassination of Benazir Bhutto as well as failing to provide her proper security. On February 12, 2011, an Anti-Terrorism Court in Rawalpindi issued an arrest warrant for Musharraf, claiming he was aware of an impending assassination attempt by the Taliban, but did not pass the information on to those responsible for protecting Bhutto.

7

Jacqueline Kennedy Onassis

Jacqueline Lee Bouvier Kennedy Onassis (July 28, 1929 – May 19, 1994) was the wife of the 35th President of the United States, John F. Kennedy, and served as First Lady of the United States during his presidency from 1961 until his assassination in 1963. Five years later she married Greek shipping magnate Aristotle Onassis; they remained married until his death in 1975. For the final two decades of her life, Jacqueline Kennedy Onassis had a successful career as a book editor. She is remembered for her contributions to the arts and preservation of historic architecture, her style, elegance, and grace. A fashion icon, her famous pink Chanel suit has become a symbol of her husband's assassination and one of the lasting images of the 1960s.

Early Life

Jacqueline Lee Bouvier was born in Southampton, New York, to Wall Street stock broker John Vernou Bouvier III (also known as "Black Jack Bouvier") and Janet Norton Lee. Jacqueline had a younger sister, Caroline Lee (known as Lee), born in 1933. Her parents divorced in 1940 and her mother married Standard Oil heir Hugh D. Auchincloss, Jr. in 1942. Through Janet's second marriage, Jacqueline gained a half sister and a half brother, Janet and James Auchincloss. Her mother's family, the Lees, were of Irish descent, and her father descended from French and English ancestors. Her maternal great grandfather emigrated from Cork, Ireland and later became the Superintendent of the New York City Public Schools. Michel Bouvier, Jacqueline's paternal great-great-grandfather, was born in France and was a contemporary of Joseph Bonaparte and Stephen Girard. He was a Philadelphia-based cabinetmaker, carpenter, merchant and real estate speculator. Michel's wife, Louise Vernou was the daughter of John Vernou, a French émigré tobacconist and Elizabeth Clifford Lindsay, an American-born woman. Jacqueline's grandfather, John Vernou Bouvier Jr., fashioned a more noble ancestry for his family in his vanity family history book *Our Forebears*. Recent scholarship and the research done by Jacqueline's cousin, John H. Davis, in his book *The Bouviers: Portrait of an American Family*, have disproved most of these fantasy lineages. She spent her early years in New York City and East Hampton, New York at the Bouvier family estate, "Lasata". Following their parents' divorce, Jacqueline and Lee divided their time between their mother's homes in McLean, Virginia and Newport, Rhode Island and their father's homes in New York City and Long Island. She attended the Chapin School in New York City. At a very early age she became an enthusiastic equestrienne, and horse-riding remained a lifelong passion.

Education and Young Adulthood

Bouvier attended the Holton-Arms School, located in Bethesda, Maryland, from 1942 to 1944 and Miss Porter's School, located in Farmington, Connecticut, from 1944 to 1947. When she made her society debut in 1947, Hearst columnist Igor Cassini dubbed her "debutante of the year'. Beginning in 1947, Bouvier spent her first two years of college at Vassar College, located in Poughkeepsie, New York, and then spent her junior year in France – at the University of Grenoble, located in Grenoble, and the Sorbonne, located in Paris – in a study-abroad program through Smith College, located in Northampton, Massachusetts. Upon returning home to the U.S., she transferred to The George Washington University, located in Washington, D.C., graduating in 1951 with a Bachelor of Arts degree in French literature. Bouvier's college graduation coincided with her sister's high school graduation, and the two spent the summer of 1951 on a trip through Europe. This trip was the subject of her only autobiographical book, *One Special Summer,* – co-authored with her sister, which is also the only one of her publications to feature her drawings.

Following her graduation, Bouvier was hired as "Inquiring Photographer" for *The Washington Times-Herald*. The position required her to pose witty questions to individuals chosen at random on the street and take their pictures to be published alongside selected quotations from their responses in the newspaper. During this time, she was engaged to a young stock broker, John Husted, for three months.

Kennedy Marriage and Family

Bouvier and then-U.S. Representative John Kennedy belonged to the same social circle and often attended the same functions. In May 1952, at a dinner party organized by mutual friends, they were formally introduced for the first time. The two began dating soon afterward, and their engagement was officially announced on June 25, 1953. Bouvier married Kennedy on September 12, 1953, at St. Mary's Church in Newport, Rhode Island in a Mass celebrated by Boston's Archbishop Richard Cushing. An estimated 700 guests attended the ceremony and 1,200 attended the reception that followed at Hammersmith Farm. The wedding cake was created by Plourde's Bakery in Fall River, Massachusetts. The wedding dress, now housed in the Kennedy Library in Boston, Massachusetts, and the dresses of her attendants were created by designer Ann Lowe of New York City. The newlyweds honeymooned in Acapulco, Mexico, before settling in their new home in McLean, Virginia. Kennedy suffered a miscarriage in 1955 and gave birth to a stillborn baby girl in 1956. That same year, the couple sold their estate, Hickory Hill, to Robert Kennedy and his wife Ethel Skakel Kennedy, moving to a townhouse on N Street in Georgetown. Kennedy subsequently gave birth to a second daughter, Caroline, in 1957, and a son, John, in 1960, both via Caesarian section.

Name	*Birth*	*Death*	*Notes*
Arabella Kennedy	August 23, 1956	August 23, 1956	Stillborn daughter.
Caroline Bouvier Kennedy	November 27, 1957		Married to Edwin Schlossberg; has two daugh ters and a son. She is the last surviving child of Jacqueline and John F. Kennedy.
John Fitzgerald Kennedy, Jr.	November 25, 1960	July 16, 1999	Magazine publisher and lawyer. Married to Carolyn Bessette. Both Kennedy and his wife died in a plane crash, as did Lauren Bessette, Carolyn's sister, on July 16, 1999, off Martha's

			Vineyard in a Piper Saratoga II HP piloted by Kennedy.
Patrick Bouvier Kennedy	August 7, 1963	August 9, 1963	Died from Hyaline Membrane Disease, today more commonly called Infant respiratory distress syndrome.

First Lady of the United States

Campaign for Presidency

On January 3, 1960, John Kennedy announced his candidacy for the Presidency and launched his nationwide campaign. Though she had initially intended to take an active role in the campaign, Kennedy learned that she was pregnant shortly after the beginning of the campaign. Due to her previous difficult pregnancies, Kennedy's doctor instructed her to stay at home. From Georgetown, Kennedy participated in her husband's campaign by answering letters, taping television commercials, giving televised and printed interviews, and writing a weekly syndicated newspaper column, "Campaign Wife." She made rare personal appearances.

As First Lady

In the general election on November 8, 1960, John F. Kennedy narrowly beat Republican Richard Nixon in the U.S. presidential election. A little over two weeks later, Jacqueline Kennedy gave birth to the couple's first son, John, Jr. When her husband was sworn in as president on January 20, 1961, Kennedy became, at age 31, one of the youngest First Ladies in history, behind Frances Folsom Cleveland and Julia Tyler. Like any First Lady, Kennedy was thrust into the spotlight and while she did not mind giving interviews or being photographed, she preferred to maintain as much privacy as possible for herself and her children. Kennedy is remembered for reorganizing entertainment for White House social events, restoring the interior of the presidential home, her taste in clothing worn during her husband's presidency, her popularity among foreign dignitaries, and leading the country in mourning after JFK's 1963 assassination. Kennedy ranks among the most popular of First Ladies.

Death of Youngest Son

Early in 1963, Kennedy became pregnant again and

curtailed her official duties. She spent most of the summer at the Kennedys' rented home on Squaw Island, near the Kennedy family's Cape Cod compound at Hyannis Port, where she went into premature labor on August 7, 1963. She gave birth to a boy, Patrick Bouvier Kennedy, via emergency Caesarian section at Otis Air Force Base, five and a half weeks prematurely. His lungs were not fully developed, and he died at Boston Children's Hospital of hyaline membrane disease (now known as respiratory distress syndrome) on August 9, 1963. The couple was devastated by the loss of their infant son, but that tragedy brought them closer together than ever before.

Death

In January 1994 Onassis was diagnosed with non-Hodgkin's lymphoma, a form of cancer. Her diagnosis was announced to the public the following month. The family and doctors were initially optimistic, and she stopped smoking at the insistence of her daughter. Onassis continued her work with Doubleday, but curtailed her schedule. By April, the cancer had spread, and she made her last trip home from New York Hospital-Cornell Medical Center on May 18, 1994. A large crowd of well-wishers, tourists, and reporters gathered on the street outside her apartment. Onassis died in her sleep at 10:15 p.m. on Thursday, May 19, two and a half months before her 65th birthday. In announcing her death, Jacqueline's son, John Kennedy Jr., stated, "My mother died surrounded by her friends and her family and her books, and the people and the things that she loved.

8

Cleopatra V.

Cleopatra V Tryphaena of Egypt (born c. 95 BC, died c. 69/68 BC or c. 57 BC) was a Ptolemaic Queen of Egypt. She is the only surely attested wife of Ptolemy XII.

Descent and Marriage

Cleopatra V Tryphaena may have been an illegitimate daughter of Ptolemy IX or she may have been the daughter of Ptolemy X. In some sources Cleopatra Tryphaena, wife of Ptolemy XII is referred to as Cleopatra VI. She is first mentioned in 79 BC in two papyri. In that year she married Ptolemy XII, king of Egypt. Ptolemy XII was an illegitimate child of Ptolemy IX, but it is unclear if he and Cleopatra Tryphaena were full siblings or not. They received divine worship as *theoí Philopátores kai Philádelphoi* (*father-, brother- and sister loving gods*). In another theory she is likely to be a daughter of Ptolemy X Alexander because Cleopatra VII reveals her grandfather as Ptolemy X Alexander and that Strabo confirms her sister Berenice's mother as Cleopatra V. In this case her mother could have been Berenice III.

Death and Identity

It is unclear how long Cleopatra V lived, and with which mentions of Cleopatra Tryphaena in the historical record she should be identified (the numbering used to distinguish the Ptolemies is a modern invention). Cleopatra Tryphaena V disappears around the same time Cleopatra Thea VII (Cleopatra VII) is born (69 B.C), she may have died from childbirth or got murdered. There is some indication that she may have died in 69 or 68 BC, because her name begins to disappear from monuments and papyri, and there is an inscription of Ptolemy

XII from 68 BC that does not mention her, when it would be expected to do so if she were still alive. If this is the case, then the Cleopatra Tryphaena, who is mentioned — after the expulsion of Ptolemy XII — as co-ruler of Egypt (together with Berenice IV) in 58 and 57 BC, and dying around 57 BC, must be her daughter, numbered by some historians as Cleopatra VI Tryphaena (this is also what Porphyry reports).

On the other hand, there is a dedication on the Temple of Edfu from 57 BC that inscribes Cleopatra Tryphaena's name alongside Ptolemy XII's, which would have meant his wife rather than daughter, and would be unlikely had Ptolemy XII's wife died 12 years earlier, even with the slow speed of news travel. Thus most modern historians consider the purported Cleopatra VI Tryphaena to be identical with Cleopatra V, and have her living to c. 57 BC. This also comports better with the account of Strabo, who reports Ptolemy XII to have had only three daughters; since we can reliably identify Berenice IV, Cleopatra VII, and Arsinoe IV as his daughters, this leaves no room for a Cleopatra VI. The historian Werner Huß believes that Ptolemy XII repudiated his wife Cleopatra V in 69 BC and married a noble Egyptian woman from the high priest family of Memphis. This presumed second wife of the Egyptian king could have been the mother of Cleopatra VII and her younger siblings, while Berenice IV was the daughter of Cleopatra V because Strabo only calls the oldest daughter of Ptolemy XII a legitimate child. If this theory is true then Cleopatra V assumed power together with her daughter Berenice IV after the expulsion of Ptolemy XII (58 BC) and died before the end of the next year, because her name again disappears after 57 BC from the documents.

9

Marilyn Monroe

Marilyn Monroe (1926-1962) Model, actress, singer and arguably one of the most famous women of the twentieth century. Monroe was born, Norma Jeane Mortenson, in June 1926. Her father was unknown and she was baptized as Norma Jeane Baker; she spent many years in foster homes because of her family situation. Monroe married Jimmy Dougherty, in 1942. When he left to the South Pacific to fight in the Second world War, she joined a local munitions factory in Burbank, California. It was here that Marilyn got her first big break. Photographer David Conover, was covering the munitions factory to show women at work. He was struck by the beauty and photogenic nature of Norma, and he used her in many of her shots. This enabled her to start a career as a model and she was soon featured on the front of many magazine covers. 1946 was a pivotal year for Marilyn, she divorced her young husband and changed her name from the boring Norma Baker to the more glamorous Marilyn Monroe. She took drama lessons and got her first movie contract with Twentieth Century Fox. Her first few films were low key, but, it gained her more prominent roles in films such as *All About Eve*, *Niagara* and later *Gentleman Prefer Blondes* and *How To Marry A Millionaire*. By now these film roles had thrust her into the global limelight. She was an iconic figure of Hollywood glamour and fashion. She was an epitome of sensuality, beauty and effervescence and was naturally photogenic. In 1954, she married baseball star Joe DiMaggio, a friend of over 2 years. They were later to divorce, but they remained close friends. She tried to move beyond the 'blonde bombshell' typecasting and set up her own movie production. She was awarded a golden globe award for her role in '*Some Like It Hot*' Tragically, she died early from an overdose of barbiturates in 1962 aged just 36.

10

Mother Teresa

Mother Teresa, winner of the Nobel Peace Prize, attained world wide fame for her life dedicated to serving the poor and destitute.

> "It is not how much we do, but how much love we put in the doing. It is not how much we give, but how much love we put in the giving."
>
> —*Mother Teresa*

Mother Teresa was born (1910) in Skopje, capital of the Republic of Macedonia. Little is known about her early life, but at a young age she felt a calling to serve through helping the poor. At the age of 18 she was given permission to join a group of nuns in Ireland. After a few months of training Mother Teresa travelled to Calcutta, India where she formally accepted the vows of a nun. In her early years she worked as a teacher in the slums of Calcutta, the widespread poverty made a deep impression on her and this led to her starting a new order called "The Missionaries of Charity". The primary objective of this mission was to look after people, who nobody else was prepared to. Mother Teresa felt that serving others was a key principle of the teachings of Jesus Christ. She often mentioned the saying of Jesus, "*Whatever you do to the least of my brethren, you do it to me.*"

> "Love cannot remain by itself — it has no meaning. Love has to be put into action, and that action is service ."
>
> —*Mother Teresa*

The Missionaries of Charity now has branches throughout the world including branches in the developed world where they work with the homeless and people affected with AIDS.

In 1965, the Society became an International Religious Family by a decree of Pope Paul VI. In the 1960s, the life of Mother Teresa was first brought to a wider public attention by Malcolm Muggeridge who wrote a book and produced a documentary called "*Something Beautiful for God*".

Awards Mother Teresa

The first Pope John XXIII Peace Prize. (1971)

- Kennedy Prize (1971)
- The Nehru Prize –"for promotion of international peace and understanding"(1972)
- Albert Schweitzer International Prize (1975),
- The Nobel Peace Prize (1979)
- States Presidential Medal of Freedom (1985)
- Congressional Gold Medal (1994)
- Honorary citizenship of the United States (November 16, 1996).

Mother Teresa was awarded the Nobel Prize "for work undertaken in the struggle to overcome poverty and distress, which also constitute a threat to peace." She refused the conventional ceremonial banquet given to laureates, and asked that the $6,000 funds be given to the poor in Calcutta. When Mother Teresa received the prize, she was asked, "What can we do to promote world peace?" Her answer was simple: "Go home and love your family." Over the last two decades of her life, Mother Teresa suffered various health problems but nothing could dissuade her from fulfilling her mission of serving the poor and needy. Until her very last illness she was active in travelling around the world to the different branches of "The Missionaries of Charity". Following Mother Teresa's death the Vatican began the process of beatification, which is the second step on the way to canonisation and sainthood. Mother Teresa was formally beatified in October 2003 by Pope John Paul II and is now known as Blessed Teresa of Calcutta. Mother Teresa was a living saint who offered a great example and inspiration to the world.

11
Anne Frank

> "It's difficult in times like these; ideals, dreams and cherished hopes rise within us, only to be crushed by grim reality. It's a wonder I haven't abandoned all my ideals, they seem so absurd and impractical. Yet I cling to them because I believe, in spite of everything, that people are truly good at heart."
>
> —*Anne Frank 21 July 1944*

Anne Frank could easily have just become another statistic in the holocaust of the Second World War. But after the war, her father Otto Frank, discovered his daughters diary. Struck by her maturity and depth of feeling, he published her diary - originally under the title *'Diary of A Young Girl'*. It became one of the most famous records of the holocaust, and helped to give a human story behind the dreadful holocaust statistics. Anne Frank was born on 12 June 1929 in Frankfurt, Germany, In 1933 (the same year as the Nazi's rise to power) her family moved to Holland, where her father ran a successful business. However, after the fall of Holland to the Nazi's in 1940, the Jewish population experienced ever increasingly repressive measures.

> "After May 1940...the trouble started for the Jews. Our freedom was severely restricted by a series of anti-Jewish decrees: Jews were required to wear a yellow star; Jews were required to turn-in their bicycles; Jews were forbidden to ride trams or in cars, even their own...Jews were forbidden to go to theatres, cinemas or any other forms of entertainment; Jews were forbidden to use swimming pools, tennis courts, hockey fields or any other athletic fields...You couldn't do this and you couldn't do that, but life went on..."
>
> —*Anne Frank 20 June 1942*

Finally, to escape arrest, Otto Frank took his family into forced hiding, behind one of his business premises in the heart of Amsterdam. Her family were later joined by the Van Pels family who were also trying to avoid arrest. Anne's diary tells of the difficulties of living in a confined space with so many people. The atmosphere was at times suffocating, but despite the hardships and difficulties of her situation, she also expressed her optimism and positive view of life and a natural joie de vivre.

> I long to ride a bike, dance, whistle, look at the world, feel young and know that I'm free, and yet I can't let it show. Just imagine what would happen if all eight of us were to feel sorry for ourselves or walk around with the discontent clearly visible on our faces. Where would that get us? (December 24, 1943)
>
> —*Anne Frank*

Unfortunately, on August 4th, 1944 (with the allies closing in on a retreating Germany army), an anonymous source gave a tip off to the German secret police. The families were arrested and sent on the last convoy train to Auschwitz. After surviving the selection process (most people under 15 were sent straight to the Gas Chambers) Anne was selected to be sent to Bergen Belsen concentration camp. It was here that Anne contracted typhoid fever and she died in March 1945, just one month before the camp was liberated by the advancing allied armies. Except her father Otto, all her family died in various concentration camps. After the war, Otto returned to the place where they had hidden for two years. It was here that he found Anne's diary and he decided to try and get it published. Her diary was published in 1947 and, following a glowing article by Jan Romein, in the newspaper Het Parool, became a best seller with people fascinated by her writing and what she managed to convey in the most difficult of situations. Her book has become an important symbol of how innocent people can suffer from intolerance and persecution.

12

Sappho

Sappho lived around 600BC on her native island of Lesbos - a large Greek island in the Aegean sea. She lived a privileged life in the aristocratic circles of Lesbos. Before being exiled along with the rest of her society, she led a group of young woman who met to discuss literature, social skills and religious knowledge. It was kind of an informal finishing school of the day. Sappho is most famous for her vast body of literature, most of which was later lost. She wrote many passionate poems, often dedicated to Aphrodite, the Greek goddess of love, sexual desire and beauty. Her poems also reflect a homoerotic bond with her young female admirers. Sappho is widely regarded as the pre-eminent pioneer of Lesbian literature. This reputation caused her works to be ignored in later times when her frank expression of sexuality was not welcomed. One legend claims Sappho died by throwing herself off a rock in despair over a lost love for a sailor called Phaon. However, like many details of her life it is difficult to be sure.

13

Oprah Winfrey

"The biggest adventure you can ever take is to live the life of your dreams."

—*Oprah Winfrey*

Oprah Winfrey was born in Kosciusko, Mississippi. Her parents were unmarried and broke up soon after conception. Oprah had a difficult childhood. She lived in great poverty and often had to dress in potato sacks for which she was mocked at school. She was also sexually abused at an early age.

"Turn your wounds into wisdom."

—*Oprah Winfrey*

From the age of 14 she went to live with her father. Oprah says he was strict, but she was in the mood to be disobedient during her teenage years. After working her way through college she became interested in journalism and media and got her first job as a news anchor for a local TV station. Her emotional style did not go down well for a news programme so she was transferred to an ailing daytime chat programme. After Oprah took over, the daily chat show took off and this later led to her own programme - The Oprah Winfrey Show. The Oprah Winfrey show has proved to be one of the most successful and highly watched TV programme of all time. It has broken many social and cultural barriers such as gay and lesbian issues. Oprah has also remained a powerful role model for women, and black American women in particular. In recent years, the Oprah Winfrey show has focused on issues of self-improvement, spirituality and self-help. Diet has also been a big issue with Oprah once successfully losing a lot of excess weight. Her subsequent diet book sold millions of copies. However, like many dieters Oprah has suffered from a yo-yo weight loss and gain. The Oprah Winfrey book club has become the most influential book clubs in the world. A recommendation from Oprah Winfrey frequently sends books to the top of the best seller lists. Many commentators agree that Oprah Winfrey exerts enormous influence. Some estimated her support for Barack Obama helped him gain 1 million votes in the 2008 election. As *Vanity Fair* said of Oprah Winfrey

> "Oprah Winfrey arguably has more influence on the culture than any university president, politician, or religious leader, except perhaps the Pope"

Oprah Winfrey was also nominated for an Oscar in the film - *A Color Purple*. Produced by Steven Spielberg, the epic *Color Purple* told of segregation in America's deep south. Oprah was widely admired for her role as Sofia.

14

Billie Jean King

Billie Jean King was a champion American tennis player winning a total of 39 grand slam titles in an illustrious career. She also played a key role in fighting for greater equality between men and women's tennis. Of her 39 grand slam titles 20 were achieved at Wimbledon. She had an aggressive, impatient style. She hit the ball very hard and was quick to come to the net. It was this style of tennis that perfectly suited the Wimbledon grass courts. One of her great rivals, Chris Evert said that her main weakness was her impatience. She first appeared at Wimbledon in 1961 as a young tennis player by the name of Billie Jean Moffat. She went on to play at Wimbledon on 22 occasions over a period of 23 years. She was a firm favourite of the crowd and in the early days was better known at Wimbledon than in her home country of US. She played a total of 265 matches at Wimbledon in both singles, doubles and mixed doubles. In 1974, she took part in one of tennis' most famous encounters - dubbed the battle of the sexes. Bobby Riggs was a former number one tennis player. Now 55 he boasted that the men's game was so superior to the women's game that he could easily beat any of the best women players of the time. Initially Billie Jean King was fearful of playing because she felt if she lost it would put women's tennis back 50 years.

> "I thought it would set us back 50 years if I didn't win that match. It would ruin the women's tour and affect all women's self esteem."

However, after Bobby Riggs beat Margaret Court (who was a great opponent of Billie Jean King and beat her twice in grand slam final) Billie Jean King took up the challenge at

Houston Astrodome in Texas. The game was watched by a crowd of over 30,000 and up to 50 million on TV. It was a media sensation and Billie Jean King proved the winner beating Bobby Riggs 6-4, 6-3, 6-3

> "Tennis is a perfect combination of violent action taking place in an atmosphere of total tranquility."
>
> —*Billie Jean King*

Throughout her career Billie Jean King campaigned for better pay and recognition for female tennis players. In her early days she was particularly critical of the US tennis authorities for their promotion of shamateurism. Billie Jean was a keen advocate of professional tennis. For her efforts related to elevating tennis she was ranked No. 5 on Sports Illustrated's "Top 40 Athletes" list for significantly altering or elevating sports the last four decades (1994). She was also named as one of the 100 most influential Americans of the 20th Century by Time Magazine. Since retirement she has worked for GBLT and an active promoter of Gay and Lesbian rights in America. She also serves on the Women's sport foundation and the Elton John AIDS campaign. In 2007, she launched Green Slam an organisation trying to make tennis more environmentally aware.

15

Eleanor of Aquitaine

Eleanor of Aquitaine (1122 or 1124 – 1 April 1204) was one of the wealthiest and most powerful women in Western Europe during the High Middle Ages. As well as being Duchess of Aquitaine in her own right, she was queen consort of France (1137–1152) and of England (1154–1189). Eleanor of Aquitaine is the only woman to have been queen of both France and England. She was the patroness of such literary figures as Wace, Benoît de Sainte-More, and Chrétien de Troyes. Eleanor succeeded her father as *suo jure* Duchess of Aquitaine and Countess of Poitiers at the age of fifteen, and thus became the most eligible bride in Europe. Three months after her accession she married Louis VII, son and junior co-ruler of her guardian, King Louis VI of France. As Queen of France, she participated in the unsuccessful Second Crusade. Soon after the Crusade was over, Louis VII and Eleanor agreed to dissolve their marriage, because of Eleanor's own desire for divorce and also because the only children they had were two daughters – Marie and Alix. The royal marriage was annulled on 11 March 1152, on the grounds of consanguinity within the fourth degree. Their daughters were declared legitimate and custody of them awarded to Louis, while Eleanor's lands were restored to her.

As soon as she arrived in Poitiers, Eleanor became engaged to Henry II, Duke of the Normans, her cousin within the third degree, who was nine years younger. On 18 May 1152, eight weeks after the annulment of her first marriage, Eleanor married the Duke of the Normans. On 25 October 1154 her husband ascended the throne of the Kingdom of England, making Eleanor Queen of the English. Over the next thirteen years, she bore Henry eight children: five sons, three of whom would become king, and three daughters. However, Henry and Eleanor eventually became estranged. She was imprisoned between 1173 and 1189 for supporting her son Henry's revolt against her husband, King Henry II. Eleanor was widowed on 6 July 1189. Her husband was succeeded by their son, Richard the Lionheart, who immediately moved to release his mother. Now queen dowager, Eleanor acted as a regent for her son while he went off on the Third Crusade. Eleanor survived her son Richard and lived well into the reign of her youngest son King John. By the time of her death she had outlived all of her children except for King John and Eleanor, Queen of Castile.

Early Life

The exact date and place of Eleanor's birth are not known. A late 13th century genealogy of her family listed her as 13 years old in the spring of 1137. Some chronicles mentionned a fidelity oath of some lords of Aquitaine on the occasion of Eleanor's fourteenth birthday in 1136. Her parents almost certainly married in 1121. Her birth place may have been Poitiers, Bordeaux, or Nieul-sur-l'Autise, where her mother died as she was 6 or 8. Eleanor or Aliénor was the oldest of three children of William X, Duke of Aquitaine, whose glittering ducal court was on the leading edge of early–12th-century culture, and his wife, Aenor de Châtellerault, the daughter of Aimeric I, Viscount of Châtellerault, and Dangereuse, who was William IX's longtime mistress as well as Eleanor's maternal grandmother. Her parents' marriage had been arranged by Dangereuse with her paternal grandfather, the Troubadour. Eleanor was named for her mother Aenor and called *Aliénor*, from the Latin *alia Aenor*, which means *the other Aenor*. It became *Eléanor* in the *langues d'oïl* (Northern French) and *Eleanor* in English. There is, however, an earlier Eleanor on record: Eleanor of Normandy, William the Conqueror's aunt, who lived a century earlier than Eleanor of Aquitaine.

By all accounts, Eleanor's father ensured that she had the best possible education. Although her native tongue was Poitevin, she was taught to read and speak Latin, was well versed in music and literature, and schooled in riding, hawking, and hunting. Eleanor was extroverted, lively, intelligent, and strong willed. In the spring of 1130, when Eleanor was six, her four-year-old brother William Aigret and their mother died at the castle of Talmont, on Aquitaine's Atlantic coast. Eleanor became the heir presumptive to her father's domains. The Duchy of Aquitaine was the largest and richest province of France; Poitou (where Eleanor spent most of her childhood) and Aquitaine together were almost one-third the size of modern France. Eleanor had only one other legitimate sibling, a younger sister named Aelith but always called Petronilla. Her half brothers, William and Joscelin, were acknowledged by William X as his sons, but not as his heirs. Later, during the first four years of Henry II's reign, all three siblings joined Eleanor's royal household.

In 1138, Duke William X set out from Poitiers to Bordeaux, taking his daughters with him. Upon reaching Bordeaux, he left Eleanor and Petronilla in the charge of the Archbishop of Bordeaux, one of the Duke's few loyal vassals who could be entrusted with the safety of the duke's daughters. The duke then set out for the Shrine of Saint James of Compostela, in the company of other pilgrims; however, he died on Good Friday 9 April 1137. Eleanor, aged about fifteen, became the Duchess of Aquitaine, and thus the most eligible heiress in Europe. As these were the days when kidnapping an heiress was seen as a viable option for obtaining a title, William had dictated a will on the very day he died, bequeathing his domains to Eleanor and appointing King Louis VI of France as her guardian. The King of France himself was also gravely ill at that time, suffering "a flux of the bowels" (dysentery) from which he seemed unlikely to recover. Despite his immense obesity and impending mortality, however, Louis the Fat remained clear-minded. To his concerns regarding his new heir, Louis, who had been destined for the monastic life of a younger son (the former heir, Philip, having died from a riding accident), was added joy over the death of one of his most powerful vassals – and the availability of the best duchy in France.

Appearance

Contemporary sources praise Eleanor's beauty. Even in

an era when ladies of the nobility were excessively praised, their praise of her was undoubtedly sincere. When she was young, she was described as *perpulchra* – more than beautiful. When she was around 30, Bernard de Ventadour, a noted troubadour, called her "gracious, lovely, the embodiment of charm," extolling her "lovely eyes and noble countenance" and declaring that she was "one meet to crown the state of any king." William of Newburgh emphasized the charms of her person, and even in her old age, Richard of Devizes described her as beautiful, while Matthew Paris, writing in the 13th century, recalled her "admirable beauty." However, no one left a more detailed description of Eleanor; the color of her hair and eyes, for example are unknown. The effigy on her tomb shows a tall and large-boned woman, though this may not be an accurate representation. Her seal of c. 1152 shows a woman with a slender figure, but this is likely an impersonal image.

In Historical Fiction

Eleanor and Henry are the main characters in James Goldman's play *The Lion in Winter*, which was made into a film starring Peter O'Toole and Katharine Hepburn in 1968 (for which Hepburn won the Academy Award for Best Actress and the BAFTA Award for Best Actress in a Leading Role and was nominated for the Golden Globe Award for Best Actress - Motion Picture Drama), and remade for television in 2003 with Patrick Stewart and Glenn Close (for which Close won the Golden Globe Award for Best Performance by an Actress In A Mini-series or Motion Picture Made for Television and was nominated for the Primetime Emmy Award for Outstanding Lead Actress - Miniseries or a Movie).

The depiction of Eleanor in the play *Becket*, which was filmed in 1964 with Pamela Brown as Eleanor, contains historical inaccuracies, as acknowledged by the author, Jean Anouilh. In 2004, Catherine Muschamp's one-woman play, *Mother of the Pride*, toured the UK with Eileen Page in the title role. In 2005, Chapelle Jaffe played the same part in Toronto. Eleanor is also associated with Nicole des Jardins in Arthur C. Clark's series *Rendezvous with Rama*. She is often cited as a role model for Nicole along with Joan of Arc

16

Hillary Rodham Clinton

Hillary Diane Rodham Clinton (born October 26, 1947) is the 67th United States Secretary of State, serving in the administration of President Barack Obama. She was a United States Senator for New York from 2001 to 2009. As the wife of the 42nd President of the United States, Bill Clinton, she was the First Lady of the United States from 1993 to 2001. In the 2008 election, Clinton was a leading candidate for the Democratic presidential nomination. A native of Illinois, Hillary Rodham first attracted national attention in 1969 for her remarks as the first student commencement speaker at Wellesley College. She embarked on a career in law after graduating from Yale Law School in 1973. Following a stint as a Congressional legal counsel, she moved to Arkansas in 1974 and married Bill Clinton in 1975. Rodham cofounded the Arkansas Advocates for Children and Families in 1977 and became the first female chair of the Legal Services Corporation in 1978. Named the first female partner at Rose Law Firm in 1979, she was twice listed as one of the 100 most influential lawyers in America. First Lady of Arkansas from 1979 to 1981 and 1983 to 1992 with husband Bill as Governor, she successfully led a task force to reform Arkansas's education system. She sat on the board of directors of Wal-Mart and several other corporations.

In 1994 as First Lady of the United States, her major initiative, the Clinton health care plan, failed to gain approval from the U.S. Congress. However, in 1997 and 1999, Clinton played a role in advocating the creation of the State Children's Health Insurance Program, the Adoption and Safe Families Act, and the Foster Care Independence Act. Her years as

First Lady drew a polarized response from the American public. The only First Lady to have been subpoenaed, she testified before a federal grand jury in 1996 due to the Whitewater controversy, but was never charged with wrongdoing in this or several other investigations during her husband's administration. The state of her marriage was the subject of considerable speculation following the Lewinsky scandal in 1998. After moving to the state of New York, Clinton was elected as a U.S. Senator in 2000. That election marked the first time an American First Lady had run for public office; Clinton was also the first female senator to represent the state. In the Senate, she initially supported the Bush administration on some foreign policy issues, including a vote for the Iraq War Resolution. She subsequently opposed the administration on its conduct of the war in Iraq and on most domestic issues. Senator Clinton was reelected by a wide margin in 2006. In the 2008 presidential nomination race, Hillary Clinton won more primaries and delegates than any other female candidate in American history, but narrowly lost to Illinois Senator Barack Obama. As Secretary of State, Clinton became the first former First Lady to serve in a president's cabinet. She has put into place institutional changes seeking to maximize departmental effectiveness and promote the empowerment of women worldwide. She has set records for most-traveled secretary for time in office. She has been at the forefront of the U.S. response to the 2011 Middle East protests, including advocating for the military intervention in Libya.

Early Life

Hillary Diane Rodham was born at Edgewater Hospital in Chicago, Illinois. She was raised in a United Methodist family, first in Chicago and then, from the age of three, in suburban Park Ridge, Illinois. Her father, Hugh Ellsworth Rodham, was the son of Welsh and English immigrants; he managed a successful small business in the textile industry. Her mother, Dorothy Emma Howell, is a homemaker of English, Scottish, French, French Canadian, and Welsh descent. She has two younger brothers, Hugh and Tony. As a child, Hillary Rodham was a teacher's favorite at her public schools in Park Ridge. She participated in swimming, baseball, and other sports. She also earned numerous awards as a Brownie and Girl Scout. She attended Maine East High School, where she participated in student council, the school newspaper, and was selected for National Honor Society. For her senior year, she was redistricted to Maine South High School, where she was a National Merit Finalist and graduated in the top five percent of her class of 1965. Her mother wanted her to have an independent, professional career, and her father, otherwise a traditionalist, held the modern notion that his daughter's abilities and opportunities should not be limited by gender. Raised in a politically conservative household, at age thirteen Rodham helped canvass South Side Chicago following the very close 1960 U.S. presidential election, where she found evidence of electoral fraud against Republican candidate Richard Nixon. She then volunteered to campaign for Republican candidate Barry Goldwater in the U.S. presidential election of 1964. Rodham's early political development was shaped most by her high school history teacher (like her father, a fervent anticommunist), who introduced her to Goldwater's classic *The Conscience of a Conservative*, and by her Methodist youth minister (like her mother, concerned with issues of social justice), with whom she saw and met civil rights leader Martin Luther King, Jr., in Chicago in 1962.

First Lady of the United States

Role as First Lady

When Bill Clinton took office as president in January 1993, Hillary Rodham Clinton became the First Lady of the United States, and announced that she would be using that form of her name. She was the first First Lady to hold a postgraduate degree and to have her own professional career up to the time of entering the White House. She was also the first to have an office in the West Wing of the White House in addition to the usual First Lady offices in the East Wing. She was part of the innermost circle vetting appointments to the new administration, and her choices filled at least eleven top-level positions and dozens more lower-level ones. She is regarded as the most openly empowered presidential wife in American history, save for Eleanor Roosevelt.

Some critics called it inappropriate for the First Lady to

play a central role in matters of public policy. Supporters pointed out that Clinton's role in policy was no different from that of other White House advisors and that voters were well aware that she would play an active role in her husband's presidency. Bill Clinton's campaign promise of "two for the price of one" led opponents to refer derisively to the Clintons as "co-presidents", or sometimes the Arkansas label "Billary". The pressures of conflicting ideas about the role of a First Lady were enough to send Clinton into "imaginary discussions" with the also-politically-active Eleanor Roosevelt. from the time she came to Washington, she also found refuge in a prayer group of The Fellowship that featured many wives of conservative Washington figures. Triggered in part by the death of her father in April 1993, she publicly sought to find a synthesis of Methodist teachings, liberal religious political philosophy, and *Tikkun* editor Michael Lerner's "politics of meaning" to overcome what she saw as America's "sleeping sickness of the soul" and that would lead to a willingness "to remold society by redefining what it means to be a human being in the twentieth century, moving into a new millennium." Other segments of the public focused on her appearance, which had evolved over time from inattention to fashion during her days in Arkansas, to a popular site in the early days of the World Wide Web devoted to showing her many different, and frequently analyzed, hairstyles as First Lady, to an appearance on the cover of *Vogue* magazine in 1998.

Senate Election of 2000

The long-serving United States Senator from New York, Daniel Patrick Moynihan, announced his retirement in November 1998. Several prominent Democratic figures, including Representative Charles B. Rangel of New York, urged Clinton to run for Moynihan's open seat in the United States Senate election of 2000. Once she decided to run, the Clintons purchased a home in Chappaqua, New York, north of New York City, in September 1999. She became the first First Lady of the United States to be a candidate for elected office.

United States Senator

Upon entering the Senate, Clinton maintained a low public

profile and built relationships with senators from both parties. She forged alliances with religiously inclined senators by becoming a regular participant in the Senate Prayer Breakfast. Clinton has served on five Senate committees: Committee on Budget (2001–2002), Committee on Armed Services (since 2003), Committee on Environment and Public Works (since 2001), Committee on Health, Education, Labor and Pensions (since 2001) and Special Committee on Aging.

Reelection Campaign of 2006

In November 2004, Clinton announced that she would seek a second Senate term. The early frontrunner for the Republican nomination, Westchester County District Attorney Jeanine Pirro, withdrew from the contest after several months of poor campaign performance. Clinton easily won the Democratic nomination over opposition from antiwar activist Jonathan Tasini. Clinton's eventual opponents in the general election were Republican candidate John Spencer, a former mayor of Yonkers, along with several third-party candidates. She won the election on November 7, 2006, with 67 percent of the vote to Spencer's 31 percent, carrying all but four of New York's sixty-two counties.

Second Term

Clinton opposed the Iraq War troop surge of 2007. In March 2007, she voted in favor of a war-spending bill that required President Bush to begin withdrawing troops from Iraq by a deadline; it passed almost completely along party lines but was subsequently vetoed by President Bush. In May 2007, a compromise war funding bill that removed withdrawal deadlines but tied funding to progress benchmarks for the Iraqi government passed the Senate by a vote of 80-14 and would be signed by Bush; Clinton was one of those who voted against it. Clinton responded to General David Petraeus's September 2007 Report to Congress on the Situation in Iraq by saying, "I think that the reports that you provide to us really require a willing suspension of disbelief."

Secretary of State

Nomination and Confirmation

In mid-November 2008, President-elect Obama and Clinton

discussed the possibility of her serving as U.S. Secretary of State in his administration, and on November 21, reports indicated that she had accepted the position. On December 1, President-elect Obama formally announced that Clinton would be his nominee for Secretary of State. Clinton said she was reluctant to leave the Senate, but that the new position represented a "difficult and exciting adventure". As part of the nomination and in order to relieve concerns of conflict of interest, Bill Clinton agreed to accept several conditions and restrictions regarding his ongoing activities and fundraising efforts for the Clinton Presidential Center and Clinton Global Initiative. The appointment required a Saxbe fix, passed and signed into law in December 2008. Confirmation hearings before the Senate Foreign Relations Committee began on January 13, 2009, a week before the Obama inauguration; two days later, the Committee voted 16–1 to approve Clinton.

Tenure

Clinton spent her initial days as Secretary of State telephoning dozens of world leaders and indicating that U.S. foreign policy would change direction: "We have a lot of damage to repair." She advocated an expanded role in global economic issues for the State Department and cited the need for an increased U.S. diplomatic presence, especially in Iraq where the Defense Department had conducted diplomatic missions.

Awards and Honors

Clinton has received many awards and honors during her career from American and international organizations for her activities concerning health, women, and children.

17

Brigitte Bardot

Brigitte Anne-Marie Bardot (born 28 September 1934) is a French former fashion model, actress and singer, and animal rights activist. In her early life, Bardot was an aspiring ballet dancer. She started her acting career in 1952 and, after appearing in 16 films, became world-famous due to her role in her then-husband Roger Vadim's controversial film *And God Created Woman*. She later starred in Jean-Luc Godard's 1963 cult film, *Contempt*. She was nominated for a BAFTA Award for Best Foreign Actress for her role in Louis Malle's 1965 film, *Viva Maria!*. She caught the attention of French intellectuals. She was the subject of Simone de Beauvoir's 1959 essay, *The Lolita Syndrome*, which described Bardot as a "locomotive of women's history" and built upon existentialist themes to declare her the first and most liberated woman of post-war France. Bardot retired from the entertainment industry in 1973.

Early Life and Career

Brigitte Bardot was born in Paris to Anne-Marie 'Toty' Mucel (1912–1978) and Louis 'Pilou' Bardot (1896–1975). Her father had an engineering degree and worked with his own father in the family business. Toty was sixteen years younger and they married in 1933. She grew up in a middle-class observant Roman Catholic family. Brigitte's mother enrolled Brigitte and her younger sister Marie-Jean (born 5 May 1938) in dance. Marie-Jean eventually gave up dancing lessons to complete her education, whereas Brigitte decided to concentrate on a ballet career. In 1947, Bardot was accepted to the Conservatoire de Paris, and for three years attended the ballet classes of Russian choreographer Boris Knyazev. (One of her classmates was Leslie Caron; fellow ballerinas nicknamed Bardot: Bichette [Little Doe]). At the invitation of her an acquaintance of her mother, she modeled in a fashion show in 1949. In the same year, she modeled for a fashion magazine "*Jardin des Modes*" managed by journalist Hélène Lazareff. Aged 15, she appeared on an 8 March 1950 cover of ELLE and was noticed by a young film director, Roger Vadim, while babysitting. He showed an issue of the magazine to director and screenwriter Marc Allégret who offered Bardot the opportunity to audition for "*Les lauriers sont coupés*" thereafter. Although Bardot got the role, the shooting of the film was cancelled but it made her consider becoming an actress. Moreover, her acquaintance with Vadim, who attended the audition, influenced her further life and career.

Although the European film industry was then in its ascendancy, Bardot was one of the few European actresses to have the mass media's attention in the United States, an interest which she did not reciprocate, rarely if ever going to Hollywood. She debuted in a 1952 comedy film *Le Trou Normand* (English title: *Crazy for Love*). From 1952-56 she appeared in seventeen films; in 1953 she played a role in Jean Anouilh's stageplay *L'Invitation au château* (*Invitation to the Castle*). She received media attention when she attended the Cannes Film Festival in April 1953. Her films of the early and mid 1950s were generally lightweight romantic dramas, some of them historical, in which she was cast as ingénue or siren, often in varying states of undress. She played bit parts in three English-language films, the British comedy *Doctor at Sea* (1955) with Dirk Bogarde, *Helen of Troy* (1954), in which she was understudy for the title role but only appears as Helen's handmaid, and *Act of Love* (1954) with Kirk Douglas. Her French-language films were dubbed for international release. Roger Vadim was not content with this light fare. The New Wave of French and Italian art directors and their stars were riding high internationally, and he felt Bardot was being undersold. Looking for something more like an art film to push her as a serious actress, he showcased her in *And God Created Woman* (1956) with Jean-Louis Trintignant. The film, about an immoral teenager in a respectable small-town setting, was an international success. In Bardot's early career, professional photographer Sam Levin's photos contributed to her image of sensuality. One of Levin's pictures shows Brigitte from behind, dressed in a white corset. British photographer Cornel Lucas made iconic images of Bardot in the 1950s and 1960s that have become representative of her public persona. She divorced Vadim in 1957 and in 1959 married actor Jacques Charrier, with whom she starred in *Babette Goes to War* in 1959. The paparazzi preyed upon her marriage, while she and her husband clashed over the direction of her career.

Vie privée (1960), directed by Louis Malle has more than an element of her life story in it. The scene in which, returning to her apartment, Bardot's character is harangued in the elevator by a middle-aged cleaning lady calling her offensive names, was based on an actual incident, and is a resonant image of celebrity in the mid-20th century. Bardot was awarded a David di Donatello Award for Best Foreign actress for the role. Soon afterwards, Bardot withdrew to the seclusion of Southern France where she had bought the house *La Madrague* in Saint-Tropez in May 1958. In 1963, she starred in Jean-Luc Godard's critically acclaimed film *Contempt*. Bardot was featured in many other films along with notable actors such as Alain Delon (*Famous Love Affairs*, *Spirits of the Dead*), Jean Gabin (*In Case of Adversity*), Sean Connery (*Shalako*), Jean Marais (*Royal Affairs in Versailles*, *School for Love*), Lino Ventura (*Rum Runners*), Annie Girardot (*The Novices*), Claudia Cardinale (*The Legend of Frenchie King*), Jeanne Moreau (*Viva Maria!*), Jane Birkin (*Don Juan, or If Don Juan Were a Woman*).

In 1973, Bardot announced that she was retiring from acting as "a way to get out elegantly". She participated in various musical shows and recorded many popular songs in the 1960s and 1970s, mostly in collaboration with Serge Gainsbourg, Bob Zagury and Sacha Distel, including "Harley Davidson", "Je Me Donne A Qui Me Plait", "Bubble gum", "Contact", "Je Reviendrais Toujours Vers Toi", "L'Appareil A Sous", "La Madrague", "On Demenage", "Sidonie", "Tu Veux, Ou Tu Veux Pas?", "Le Soleil De Ma Vie" (the cover of Stevie Wonder's "You Are the Sunshine of My Life") and the notorious

"Je t'aime... moi non plus". Bardot pleaded with Gainsbourg not to release this duet and he complied with her wishes; the following year he re-recorded a version with British-born model and actress Jane Birkin, which became a massive hit all over Europe. The version with Bardot was issued in 1986 and became a popular download hit in 2006 when Universal Records made their back catalogue available to purchase online, with this version of the song ranking as the third most popular download.

Personal Life

On 21 December 1952, aged 18, Bardot was married to director Roger Vadim. In order to receive permission from Bardot's parents to marry her, Vadim, originally a Russian Orthodox Christian, was urged to convert to Catholicism, although it is not clear if he ever did so. They divorced five years later, but remained friends and collaborated in later work. Bardot had an affair with her *And God Created Woman* co-star Jean-Louis Trintignant (married at the time to actress Stephane Audran) before her divorce from Vadim. The two lived together for about two years. Their relationship was complicated by Trintignant's frequent absence due to military service and Bardot's affair with musician Gilbert Bécaud, and they eventually separated. The 9 February 1958 edition of the *Los Angeles Times* reported on the front page that Bardot was recovering in Italy from a reported nervous breakdown. A suicide attempt with sleeping pills two days earlier was denied by her public relations manager.

On 18 June 1959, she married actor Jacques Charrier, by whom she had her only child, a son, Nicolas-Jacques Charrier (born 11 January 1960). After she and Charrier divorced in 1962, Nicolas was raised in the Charrier family and did not maintain close contact with Bardot until his adulthood. Bardot's other husbands were German millionaire playboy Gunter Sachs (14 July 1966 – 1 October 1969) and Bernard d'Ormale (16 August 1992 – present). In the late 1950s, she shared an exchange she considered "la croisée de deux sillages" ("the crossing of two wakes") with actor and true crime author John Gilmore, then an actor in France who was working on a New Wave film with Jean Seberg.

Animal Welfare Activism

In 1973, before her 39th birthday, Bardot announced her retirement. After appearing in more than forty motion pictures and recording several music albums, most notably with Serge Gainsbourg, she chose to use her fame to promote animal rights. In 1986, she established the Brigitte Bardot Foundation for the Welfare and Protection of Animals. She became a vegetarian and raised 3,000,000 francs to fund the foundation by auctioning off jewelry and many personal belongings. Today she is a strong animal rights activist and a major opponent of the consumption of horse meat.

Politics and Legal Issues

Bardot expressed support for President Charles de Gaulle in the 1960s. Her husband Bernard d'Ormale is a former adviser of the Front National, the main nationalist party in France. Bardot never joined the party and is not a known sympathiser. In a book she wrote in 1999, called "Le Carré de Pluton" (Pluto's Square), Bardot criticizes the procedure used in the ritual slaughter of sheep during the Muslim festival of Eid al-Adha. Additionally, in a section in the book entitled, *Open Letter to My Lost France*, Bardot writes: "...my country, France, my homeland, my land is again invaded by an overpopulation of foreigners, especially Muslims.". For this comment, a French court fined her 30,000 francs in June 2000. She had previously been fined in 1997 for the original publication of this open letter in Le Figaro and again 1998 for making similar remarks. In May 2003, the Movement Against Racism and for Friendship between Peoples (MRAP) announced they were going to sue Bardot for the comments. The "Ligue des droits de l'homme" (Human Rights League) announced they were considering similar legal proceedings.

In 2008, she was convicted of inciting racial/religious hatred in relation to a letter she wrote, a copy of which she sent to Nicolas Sarkozy when he was Interior Minister of France.

18

Eleanor Roosevelt

Eleanor Roosevelt was First Lady of the United States from 1933 to 1945. She was also an advocate for human and civil rights. She was made a delegate to the UN General Assembly and played a key role in in drafting the Universal declaration of human rights. She also was a key figure in John F. Kennedy's Presidential Commission on the status of women. She was born October 11, 1884. At an early age her mother and father passed away, leaving her to be brought up by her maternal grandmother. In her late teens, she became active in social work, volunteering in the slums of east New York. In 1902, she also met Franklin D.Roosevelt who was currently studying at Harvard. After some opposition from Franklin's mother, they married in 1905. They had six children, however the marriage was not without difficulties. In 1918, Eleanor discovered Franklin was having an affair. This nearly led to a divorce, but was avoided after family members and advisers encouraged them to stay together. Nevertheless, it changed the marriage, and Eleanor became more independent as a result. By 1921, Franklin was stricken with a paralyis of the legs which left him in a wheel chair. Eleanor was pivotal in helping Franklin deal with his disability and successfully encouraged him to return to public life.

In the 1920s, Eleanor worked as a teacher, teaching literature and American history in New York. However, she increasingly found time to become involved in local and state politics. She became influential in the New York Democratic party and campaigned on a variety of issues such as promoting the aims of the Women's Trade Union League. Eleanor's political activism and popularity helped her husband's political career, especially in gaining votes from women and labour organisations. In 1932, Franklin D.Roosevelt was elected President, against a backdrop of the Great Depression and mass unemployment. He was elected with a new mandate for greater government intervention in trying to offer a 'New Deal' to the unemployed. As First Lady, Eleanor broke the traditional mould and became a highly visible figure speaking at rallies and visiting unemployed workers. People felt a sincerity in her desire to aid the marginalised of the great depression and her public profile was very high. Eleanor also provided a role model for women, at a time when women rarely ventured out of a domesticated life. To articulate her beliefs, Eleanor wrote a daily newspaper column and other articles; these ranged from women's issues to general humanitarian causes.

Second World War

During the second world war, Eleanor took on even more duties and spent her time offering moral support to the war effort. In 1943, her visit to troops in the Pacific proved a great success, with even sceptics admitting her presence and determination had great effect on improving the morale of the men. During the war, Eleanor also supported the rights of black Americans (such as the Tuskegee Airmen) to train as fighter pilots and take a more visible role in the American military effort. At various times, Eleanor made efforts to seek greater civil rights for the US black population, which at the time still experienced much discrimination. In the final months of the war, Franklin passed away. The pain of Franklin's death was heightened for Eleanor by knowing he spent his final moments with another Lucy Mercer, the women he had earlier had an affair with.

United Nations

After the war, Harry Truman, who succeeded Franklin D. Roosevelt, appointed Eleanor as a delegate to the United Nations General Assembly.

> *"The mobilization of world opinion and methods of negotiation should be developed and used by every nation in order to strengthen the United Nations."*
>
> *—Eleanor Roosevelt.*

With great enthusiasm for the ideal of the United Nations, Eleanor helped draft the Universal Declaration of Human Rights, which was adopted by General Assembly on December 10, 1948. Eleanor considered this her greatest achievements and called it *'an international Magna Carta of all mankind'*. She was active in the United Nations until Eisenhower's electoral victory in 1953.

> "We stand today at the threshold of a great event both in the life of the United Nations and in the life of mankind. This declaration may well become the international Magna Carta for all men everywhere."
>
> —*Eleanor Roosevelt*

In the post war period, many encouraged her to run for public office, but she declined preferring to concentrate on non-partisan public works. After the election of John F. Kennedy, she was reappointed to the United Nations between 1961 and 1962. After developing bone marrow tuberculosis, she passed away on November 7, 1962. Her funeral was attended by presidents Truman, Eisenhower and John F. Kennedy.

19

Teresa of Ávila

Saint Teresa of Ávila, (March 28, 1515 – October 4, 1582) was a prominent Spanish mystic, Roman Catholic saint, Carmelite nun, and writer of the Counter Reformation, and theologian of contemplative life through mental prayer. She was a reformer of the Carmelite Order and is considered to be, along with John of the Cross, a founder of the Discalced Carmelites. In 1622, forty years after her death, she was canonized by Pope Gregory XV, and in 1970 named a Doctor of the Church by Pope Paul VI. Her books, which include her autobiography, *The Life of Teresa of Jesus*, and her seminal work, *El Castillo Interior* (The Interior Castle), are an integral part of the Spanish Renaissance literature as well as Christian mysticism and Christian meditation practices as she entails in her other important work *Camino de Perfección* (The Way of Perfection).She died in 1582.

Early Life

Teresa de Cepeda y Ahumada was born in 1515 in Gotarrendura, in the province of Ávila, Spain. Her paternal grandfather, *Juan de Toledo*, was a marrano (Jewish forced-convert to Christianity) and was condemned by the Spanish Inquisition for allegedly returning to the Jewish faith. Her father, Alonso Sánchez de Cepeda, bought knighthood and successfully assimilated into Christian society. Teresa's mother, Beatriz, was especially keen to raise her daughter as a pious Christian. Teresa was fascinated by accounts of the lives of the saints, and ran away from home at age seven with her brother Rodrigo to find martyrdom among the Moors. Her uncle stopped them as he was returning to the city, having spotted the two outside the city walls. In the cloister, she suffered greatly from illness. Early in her sickness, she experienced periods of religious ecstasy through the use of the devotional book *"Tercer abecedario espiritual,"* translated as the *Third Spiritual Alphabet* (published in 1527 and written by Francisco de Osuna). This work, following the example of similar writings of medieval mystics, consisted of directions for examinations of conscience and for spiritual self-concentration and inner contemplation (known in mystical nomenclature as *oratio recollectionis* or *oratio mentalis*). She also employed other mystical ascetic works such as the *Tractatus de oratione et meditatione* of Saint Peter of Alcantara, and perhaps many of those upon which Saint Ignatius of Loyola based his *Spiritual Exercises* and possibly the *Spiritual Exercises* themselves.

She claimed that during her illness she rose from the lowest stage, "recollection", to the "devotions of silence" or even to the "devotions of ecstasy", which was one of perfect union with God. During this final stage, she said she frequently experienced a rich "blessing of tears." As the Catholic distinction between mortal and venial sin became clear to her, she says she came to understand the awful terror of sin and the inherent nature of original sin. She also became conscious of her own natural impotence in confronting sin, and the necessity of absolute subjection to God. Around 1556, various friends suggested that her newfound knowledge was diabolical, not divine. She began to inflict various tortures and mortifications of the flesh upon herself. But her confessor, the Jesuit Saint Francis Borgia, reassured her of the divine inspiration of her thoughts. On St. Peter's Day in 1559, Teresa became firmly convinced that Jesus Christ presented himself to her in bodily form, though invisible. These visions lasted almost uninterrupted for more than two years.

Activities as Reformer

The incentive to give outward practical expression to her inward motive was inspired in Teresa by the Franciscan priest Saint Peter of Alcantara who became acquainted with her as Founder early in 1560, and became her spiritual guide and counselor. She now resolved to found a reformed Carmelite convent, correcting the laxity which she had found in the Cloister of the Incarnation and others. Guimara de Ulloa, a woman of wealth and a friend, supplied the funds. Teresa worked for many years encouraging the forcibly converted Jews of Spain to follow Christianity. The absolute poverty of the new monastery, established in 1562 and named St. Joseph's (San José), at first excited a scandal among the citizens and authorities of Ávila, and the little house with its chapel was in peril of suppression; but powerful patrons, including the bishop himself, as well as the impression of well-secured subsistence and prosperity, turned animosity into applause. In March 1563, when Teresa moved to the new cloister, she received the papal sanction to her prime principle of absolute poverty and renunciation of property, which she proceeded to formulate into a "Constitution". Her plan was the revival of the earlier, stricter rules, supplemented by new regulations such as the three disciplines of ceremonial flagellation prescribed for the divine service every week, and the discalceation of the nun. For the first five years, Teresa remained in pious seclusion, engaged in writing.

In 1567, she received a patent from the Carmelite general, Rubeo de Ravenna, to establish new houses of her order, and in this effort and later visitations she made long journeys through nearly all the provinces of Spain. Of these she gives a description in her *"Libro de las Fundaciones."* Between 1567 and 1571, reform convents were established at Medina del Campo, Malagon, Valladolid, Toledo, Pastrana, Salamanca, and Alba de Tormes. As part of her original patent, Teresa was given permission to set up two houses for men who wished to adopt the reforms; she convinced John of the Cross and Anthony of Jesus to help with this. They founded the first convent of Discalced Carmelite Brethren in November 1568 at Duruello. Another friend, Gerónimo Grecian, Carmelite visitator of the older observance of Andalusia and apostolic commissioner, and later provincial of the Teresian reforms, gave her powerful support in founding convents at Segovia (1571), Beas de Segura (1574), Seville (1575), and Caravaca de la Cruz (Murcia, 1576), while the deeply mystical John, by his power as teacher and preacher, promoted the inner life of the movement. In 1576 a series of persecutions began on the part of the older observant Carmelite order against Teresa, her friends, and her reforms. Pursuant to a body of resolutions adopted at the general chapter at Piacenza, the "definitors" of the order forbade all further founding of convents. The general condemned her to voluntary retirement to one of her institutions. She obeyed and chose St. Joseph's at Toledo. Her friends and subordinates were subjected to greater trials.

Finally, after several years her pleadings by letter with King Philip II of Spain secured relief. As a result, in 1579, the processes before the inquisition against her, Grecian, and others were dropped, which allowed the reform to continue. A brief of Pope Gregory XIII allowed a special provincial for the younger branch of the discalced nuns, and a royal rescript created a protective board of four assessors for the reform. During the last three years of her life, Teresa founded convents at Villanueva de la Jara in northern Andalusia (1580), Palencia (1580), Soria (1581), Burgos, and Granada (1582). In total seventeen convents, all but one founded by her, and as many men's cloisters were due to her reform activity of twenty years. Her final illness overtook her on one of her journeys from Burgos to Alba de Tormes. She died in 1582, just as Catholic nations were making the switch from the Julian to the Gregorian calendar, which required the removal of October 5–14 from the calendar.

Mysticism

The kernel of Teresa's mystical thought throughout all her writings is the ascent of the soul in four stages (*The Autobiography* Chs. 10-22):

> The first, or "mental prayer", is that of devout contemplation or concentration, the withdrawal of the soul from without and specially the devout observance of the passion of Christ and penitence (*Autobiography* 11.20).

The second is the "prayer of quiet", in which at least the human will is lost in that of God by virtue of a charismatic, supernatural state given of God, while the other faculties, such as memory, reason, and imagination, are not yet secure from worldly distraction. While a partial distraction is due to outer performances such as repetition of prayers and writing down spiritual things, yet the prevailing state is one of quietude

(*Autobiography* 14.1). The "devotion of union" is not only a supernatural but an essentially ecstatic state. Here there is also an absorption of the reason in God, and only the memory and imagination are left to ramble. This state is characterized by a blissful peace, a sweet slumber of at least the higher soul faculties, a conscious rapture in the love of God.

Major Writings

Teresa's writings, produced for didactic purposes, stand among the most remarkable in the mystical literature of the Catholic Church:

- The *"Autobiography,"* written before 1567, under the direction of her confessor, Fr Pedro Ibáñez;
- *" El Camino de Perfección"*, written also before 1567, at the direction of her confessor;
- "Meditations on Song of Songs", 1567, written nominally for her daughters at the convent of Our Lady of Mount Carmel
- *"El Castillo Interior"*, written in 1577;
- *"Relaciones"*, an extension of the autobiography giving her inner and outer experiences in epistolary form.
- Two smaller works are the *"Conceptos del Amor"* ("Concepts of Love") and *"Exclamaciones"*. In addition, there are *"Las Cartas"* (Saragossa, 1671), or her correspondence, of which there are 342 extant letters and 87 fragments of others. St Teresa's prose is marked by an unaffected grace, an ornate neatness, and charming power of expression, together placing her in the front rank of Spanish prose writers; and her rare poems (*"Todas las poesías"*, Munster, 1854) are distinguished for tenderness of feeling and rhythm of thought.

20

Agatha Christie

Dame Agatha Christie, (15 September 1890 - 12 January 1976). Agatha Christie was an English writer of crime and romantic novels. She is best remembered for her detective stories including the two diverse characters of Miss Marple and Hercule Poirot. She is considered to be the best selling writer of all time. Only the Bible is known to have outstripped her collected sales of roughly four billion world wide copies. Her works have been translated into more languages than any other individual writer. Agatha Christie was born in Torquay, Devon 1890 to Clarissa Margaret Boehmer and a wealthy American stockbroker. She was brought up by both her mother and her sister. In the First World war, she trained and worked as a nurse helping to treat wounded soldiers. She also became educated in the field of pharmacy. She recalled her time as a nurse with great fondness, saying it was one of the most rewarding jobs she ever undertook. Agatha Christie's married an aviator in the Royal Flying Corps - Archibald Christie in December 1914. The marriage was somewhat turbulent and ended in divorce in 1928, two years after Archibald had begun an affair. In 1926, Agatha Christie disappeared for 11 days. The circumstances were never really resolved and it created widespread media interest in the disappearance of this famous novelist. She was eventually discovered in a Harrogate hotel eleven days later. Though Agatha Christie never said why, it was probably a combination of shock over her mother's death and the discovery of her husband's affair. In 1930, she married her second husband, Max Mallowan. This marriage was happier, though her only child, Rosalind Hicks, came from her first marriage.

Writing Career of Agatha Christie

Agatha Christie began writing in 1920, after the end of the First World War. Her first story was *The Mysterious Affair at Styles*, (1920). This featured the soon to be famous detective - Hercule Poirot, who at the time was portrayed as a Belgian refugee from the Great War. The book sold well and helped meet the public's great appetite for detective novels. It was a genre that had been popularised through Arthur Conan Doyle's Sherlock Holmes stories at the turn of the century. Agatha Christie went on to write over 40 novels featuring the proud and immaculate Hercule Poirot. Like Conan Doyle, Christie had no great love for her own creation - Poirot seemed to be admired by the public more than the writer herself. Agatha Christie preferred her other great detective -, the quiet but effective old lady - Miss Marple. The character of Miss Marple was based on the traditional English country lady - and her own relatives.

The plot of Agatha Christies novels could be described as formulaic. Murders were committed by ingenious methods - often involving poison, which Agatha Christie had great knowledge of. After interrogating all the main suspects, the detective would bring all the participants into some drawing room before explaining who was the murderer. The psychological suspense of the novels, and the fact readers feel they have a good chance of solving the crime undoubtedly added to the popularity of the books. During the Second World War, Christie worked in the pharmacy of the University College London, which gave her ideas for some of her murder methods. After the war, her books continued to grow in international popularity. In 1952, her play *The Mousetrap* was debuted at the Ambassador's Theatre in London, and has been performed without a break ever since. Her success led to her being honoured in the New Year's honour list. In 1971 she was appointed Dame Commander of the British Empire. She died in 1976 aged 85.

21

Catherine de' Medici

Catherine de' Medici (13 April 1519 – 5 January 1589) was an Italian noblewoman who was Queen consort of France from 1547 until 1559, as the wife of King Henry II of France. In 1533, at the age of fourteen, Caterina married Henry, second son of King Francis I and Queen Claude of France. Under the gallicised version of her name, Catherine de Médicis, she was Queen consort of France as the wife of King Henry II of France from 1547 to 1559. Throughout his reign, Henry excluded Catherine from participating in state affairs and instead showered favours on his chief mistress, Diane de Poitiers, who wielded much influence over him. Henry's death thrust Catherine into the political arena as mother of the frail fifteen-year-old King Francis II. When he died in 1560, she became regent on behalf of her ten-year-old son King Charles IX and was granted sweeping powers. After Charles died in 1574, Catherine played a key role in the reign of her third son, Henry III. He dispensed with her advice only in the last months of her life.

Catherine's three sons reigned in an age of almost constant civil and religious war in France. The problems facing the monarchy were complex and daunting. At first, Catherine compromised and made concessions to the rebelling Protestants, or Huguenots, as they became known. She failed, however, to grasp the theological issues that drove their movement. Later, she resorted in frustration and anger to hard-line policies against them. In return, she came to be blamed for the excessive persecutions carried out under her sons' rule, in particular for the St. Bartholomew's Day massacre of 1572, in which thousands of Huguenots were killed in Paris and throughout

France. Some historians have excused Catherine from blame for the worst decisions of the crown, though evidence for her ruthlessness can be found in her letters. In practice, her authority was always limited by the effects of the civil wars. Her policies, therefore, may be seen as desperate measures to keep the Valois monarchy on the throne at all costs, and her patronage of the arts as an attempt to glorify a monarchy whose prestige was in steep decline. Without Catherine, it is unlikely that her sons would have remained in power. The years in which they reigned have been called "the age of Catherine de' Medici".

Marriage

On her visit to Rome, the Venetian envoy described her as "small of stature, and thin, and without delicate features, but having the protruding eyes peculiar to the Medici family". Suitors, however, lined up for her hand, including James V of Scotland who sent the Duke of Albany to Clement to conclude a marriage in April and November 1530. However, Clement jumped at the offer made by Francis I of France who proposed his second son, Henry, Duke of Orléans in early 1533. Henry was a prize catch for Catherine, who despite her wealth was from commoner origins. The wedding, a grand affair marked by extravagant display and gift-giving, took place in Marseille on 28 October 1533. Prince Henry danced and jousted for Catherine. The fourteen-year-old couple left their wedding ball at midnight to perform their nuptial duties. Henry arrived in the bedroom with King Francis, who is said to have stayed until the marriage was consummated. He noted that "each had shown valour in the joust". Clement visited the newlyweds in bed the next morning and added his blessings to the night's proceedings.

Queen Mother

Reign of Francis II

Francis II became king at the age of fifteen. In what has been called a *coup d'état*, the Cardinal of Lorraine and the Duke of Guise—whose niece, Mary, Queen of Scots, had married Francis the year before—seized power the day after Henry II's death and quickly moved themselves into the Louvre with the young couple. The English ambassador reported a few days later that "the house of Guise ruleth and doth all about the French king". For the moment, Catherine worked with the Guises out of necessity. The Guise brothers set about persecuting the Protestants with zeal. Catherine adopted a moderate stance and spoke up against the Guise persecutions, though she had no particular sympathy for the Huguenots, whose beliefs she never shared. The Protestants looked for leadership first to Antoine de Bourbon, King of Navarre, the First Prince of the Blood, and then, with more success, to his brother, Louis de Bourbon, Prince of Condé, who backed a plot to overthrow the Guises by force. When the Guises heard of the plot, they moved the court to the fortified Château of Amboise. The Duke of Guise launched an attack into the woods around the château. His troops surprised the rebels and killed many of them on the spot, including the commander, La Renaudie. Others they drowned in the river or strung up around the battlements while Catherine and the court watched. In June 1560, Michel de l'Hôpital was appointed Chancellor of France. He sought the support of France's constitutional bodies and worked closely with Catherine to defend the law in the face of the growing anarchy. Neither saw the need to punish Protestants who worshipped in private and did not take up arms. On 20 August 1560, Catherine and the chancellor advocated this policy to an assembly of notables at Fontainebleau. Historians regard the occasion as an early example of Catherine's statesmanship.

Reign of Charles IX

At first Catherine kept the nine-year-old king, who cried at his coronation, close to her, and slept in his chamber. She presided over his council, decided policy, and controlled state business and patronage. However, she was never in a position to control the country as a whole, which was on the brink of civil war. In many parts of France the rule of nobles held sway rather than that of the crown. The challenges Catherine faced were complex and in some ways difficult for her to comprehend as a foreigner. She summoned church leaders from both sides to attempt to solve their doctrinal differences. Despite her optimism, the resulting Colloquy of Poissy ended in failure on 13 October 1561, dissolving itself without her permission. Catherine failed because she saw the religious divide only in political terms. In the words of historian R. J. Knecht, "she underestimated the strength of religious conviction, imagining that all would be well if only she could get the party leaders to agree". In January 1562, Catherine issued the tolerant Edict of Saint-Germain in a further attempt to build bridges with the Protestants. On 1 March 1562, however, in an incident known as the Massacre of Vassy, the Duke of Guise and his men attacked worshipping Huguenots in a barn at Vassy (Wassy), killing 74 and wounding 104. Guise, who called the massacre "a regrettable accident", was cheered as a hero in the streets of Paris while the Huguenots called for revenge. The massacre lit the fuse that sparked the French Wars of Religion. For the next thirty years, France found itself in a state of either civil war or armed truce.

Within a month Louis de Bourbon, Prince of Condé and Admiral Gaspard de Coligny had raised an army of 1,800. They formed an alliance with England and seized town after town in France. Catherine met Coligny, but he refused to back down. She therefore told him: "Since you rely on your forces, we will show you ours". The royal army struck back quickly and laid siege to Huguenot-held Rouen. Catherine visited the deathbed of Antoine de Bourbon, King of Navarre, after he was fatally wounded by an arquebus shot. Catherine insisted on visiting the field herself and when warned of the dangers laughed, "My courage is as great as yours". The Catholics took Rouen, but their triumph was short lived. On 18 February 1563, a spy called Poltrot de Méré fired an arquebus into the back of the Duke of Guise, at the siege of Orléans. The murder triggered an aristocratic blood feud that complicated the French civil wars for years to come. Catherine, however, was delighted with the death of her ally. "If Monsieur de Guise had perished sooner", she told the Venetian ambassador, "peace would have been achieved more quickly". On 19 March 1563, the Edict of Amboise, also known as the Edict of Pacification, ended the war. Catherine now rallied both Huguenot and Catholic forces to retake Le Havre from the English.

St. Bartholomew's Day Massacre

Three days later, Admiral Coligny was walking back to

his rooms from the Louvre when a shot rang out from a house and wounded him in the hand and arm. A smoking arquebus was discovered in a window, but the culprit had made his escape from the rear of the building on a waiting horse. Coligny was carried to his lodgings at the Hôtel de Béthisy, where the surgeon Ambroise Paré removed a bullet from his elbow and amputated a damaged finger with a pair of scissors. Catherine, who was said to have received the news without emotion, made a tearful visit to Coligny and promised to punish his attacker. Many historians have blamed Catherine for the attack on Coligny. Others point to the Guise family or a Spanish-papal plot to end Coligny's influence on the king. Whatever the truth, the bloodbath that followed was soon beyond the control of Catherine or any other leader. The St. Bartholomew's Day massacre, which began two days later, has stained Catherine's reputation ever since. There is no reason to believe she was not party to the decision when on 23 August Charles IX ordered, "Then kill them all! Kill them all!". The thinking was clear. Catherine and her advisers expected a Huguenot uprising to revenge the attack on Coligny. They chose therefore to strike first and wipe out the Huguenot leaders while they were still in Paris after the wedding. The slaughter in Paris lasted for almost a week. It spread to many parts of France, where it persisted into the autumn. In the words of historian Jules Michelet, "St Bartholomew was not a day, but a season". On 29 September, when Navarre knelt before the altar as a Roman Catholic, having converted to avoid being killed, Catherine turned to the ambassadors and laughed. From this time dates the legend of the wicked Italian queen. Huguenot writers branded Catherine a scheming Italian, who had acted on Machiavelli's principles to kill all enemies in one blow.

Reign of Henry III

Two years later, Catherine faced a new crisis with the death of Charles IX at the age of twenty-three. His dying words were "oh, my mother ...". The day before he died, he named Catherine regent, since his brother and heir, Henry the Duke of Anjou, was in the Polish-Lithuanian Commonwealth, where he had been elected king the year before. However three months after his coronation at Wawel

Cathedral, Henry abandoned that throne and returned to France in order to become king of France. Catherine wrote to Henry: "I am grief-stricken to have witnessed such a scene and the love which he showed me at the end ... My only consolation is to see you here soon, as your kingdom requires, and in good health, for if I were to lose you, I would have myself buried alive with you". Henry was Catherine's favourite son. Unlike his brothers, he came to the throne as a grown man. He was also healthier, though he suffered from weak lungs and constant fatigue. His interest in the tasks of government, however, proved fitful. He depended on Catherine and her team of secretaries until the last few weeks of her life. He often hid from state affairs, immersing himself in acts of piety, such as pilgrimages and flagellation. Henry married Louise de Lorraine-Vaudémont in February 1575, two days after his coronation. His choice thwarted Catherine's plans for a political marriage to a foreign princess. Rumours of Henry's inability to produce children were by that time in wide circulation.

The papal nuncio Salviati observed, "it is only with difficulty that we can imagine there will be offspring ... physicians and those who know him well say that he has an extremely weak constitution and will not live long". As time passed and the likelihood of children from the marriage receded, Catherine's youngest son, Francis, Duke of Alençon, known as "Monsieur", played upon his role as heir to the throne, repeatedly exploiting the anarchy of the civil wars, which were by now as much about noble power struggles as religion. Catherine did all in her power to bring Francis back into the fold. On one occasion, in March 1578, she lectured him for six hours about his dangerously subversive behaviour. In 1576, in a move that endangered Henry's throne, Francis allied with the Protestant princes against the crown. On 6 May 1576, Catherine gave in to almost all Huguenot demands in the Edict of Beaulieu. The treaty became known as the *Peace of Monsieur* because it was thought that Francis had forced it on the crown. Francis died of consumption in June 1584, after a disastrous intervention in the Low Countries during which his army had been massacred. Catherine wrote, the next day: "I am so wretched to live long enough to see so many people die before me, although I realize that God's will must be obeyed, that He owns everything, and that he lends us only for as long as He likes the children whom He gives us". The death of her youngest son was a calamity for Catherine's dynastic dreams. Under Salic law, by which only males could ascend the throne, the Huguenot Henry of Navarre now became heir presumptive to the French crown. Catherine had at least taken the precaution of marrying Margaret, her youngest daughter, to Navarre. Margaret, however, became almost as much of a thorn in Catherine's side as Francis, and in 1582, she returned to the French court without her husband. Catherine was heard yelling at her for taking lovers. Catherine sent Pomponne de Bellièvre to Navarre to arrange Margaret's return. In 1585, Margaret fled Navarre again. She retreated to her property at Agen and begged her mother for money. Catherine sent her only enough "to put food on her table".

Moving on to the fortress of Carlat, Margaret took a lover called d'Aubiac. Catherine asked Henry to act before Margaret brought shame on them again. In October 1586, therefore, he had Margaret locked up in the Château d'Usson. D'Aubiac was executed, though not, despite Catherine's wish, in front of Margaret. Catherine cut Margaret out of her will and never saw her again. Catherine was unable to control Henry in the way she had Francis and Charles. Her role in his government became that of chief executive and roving diplomat. She travelled widely across the kingdom, enforcing his authority and trying to head off war. In 1578, she took on the task of pacifying the south. At the age of fifty-nine, she embarked on an eighteen-month journey around the south of France to meet Huguenot leaders face to face. Her efforts won Catherine new respect from the French people. On her return to Paris in 1579, she was greeted outside the city by the Parlement and crowds. The Venetian ambassador, Gerolamo Lipomanno, wrote: "She is an indefatigable princess, born to tame and govern a people as unruly as the French: they now recognize her merits, her concern for unity and are sorry not to have appreciated her sooner". She was under no illusions, however. On 25 November 1579, she wrote to the king, "You are on the eve of a general revolt. Anyone who tells you differently is a liar".

Catholic League

Many leading Roman Catholics were appalled by Catherine's attempts to appease the Huguenots. After the Edict of Beaulieu, they had started forming local leagues to protect their religion. The death of the heir to the throne in 1584 prompted the Duke of Guise to assume the leadership of the Catholic League. He planned to block Henry of Navarre's succession and place Henry's Catholic uncle Cardinal Charles de Bourbon on the throne instead. In this cause, he recruited the great Catholic princes, nobles and prelates, signed the treaty of Joinville with Spain, and prepared to make war on the "heretics". By 1585, Henry III had no choice but to go to war against the League. As Catherine put it, "peace is carried on a stick" (*bâton porte paix*). "Take care", she wrote to the king, "especially about your person. There is so much treachery about that I die of fear". Henry was unable to fight the Catholics and the Protestants at once, both of whom had stronger armies than his own. In the Treaty of Nemours, signed on 7 July 1585, he was forced to give in to all the League's demands, even that he pay its troops. He went into hiding to fast and pray, surrounded by a bodyguard known as "the Forty-five", and left Catherine to sort out the mess. The monarchy had lost control of the country, and was in no position to assist England in the face of the coming Spanish attack. The Spanish ambassador told Philip II that the abscess was about to burst. By 1587, the Catholic backlash against the Protestants had become a campaign across Europe. Elizabeth I of England's execution of Mary, Queen of Scots, on 18 February 1587 outraged the Catholic world. Philip II of Spain prepared for an invasion of England. The League took control of much of northern France to secure French ports for his armada.

Death

Henry hired Swiss troops to help him defend himself in Paris. The Parisians, however, claimed the right to defend the city themselves. On 12 May 1588, they set up barricades in the streets and refused to take orders from anyone except the Duke of Guise. When Catherine tried to go to mass, she found her way barred, though she was allowed through the barricades. The chronicler L'Estoile reported that she cried all through her lunch that day. She wrote to Bellièvre, "Never have I seen myself in such trouble or with so little light by which to escape". As usual, Catherine advised the king, who had fled the city in the nick of time, to compromise and live to fight another day. On 15 June 1588, Henry duly signed the Act of Union, which gave in to all the League's latest demands. On 8 September 1588 at Blois, where the court had assembled for a meeting of the Estates, Henry dismissed all his ministers without warning. Catherine, in bed with a lung infection, had been kept in the dark. The king's actions effectively ended her days of power.

Patron of the Arts

Catherine believed in the humanist ideal of the learned Renaissance prince whose authority depended on letters as well as arms. She was inspired by the example of her father-in-law, King Francis I of France, who had hosted the leading artists of Europe at his court, and by her Medici ancestors. In an age of civil war and declining respect for the monarchy, she sought to bolster royal prestige through lavish cultural display. Once in control of the royal purse, she launched a programme of artistic patronage that lasted for three decades. During this time, she presided over a distinctive late French Renaissance culture in all branches of the arts. An inventory drawn up at the Hotel de la Reine after Catherine's death shows her to have been a keen collector. Listed works of art included tapestries, hand-drawn maps, sculptures, rich fabrics, ebony furniture inlaid with ivory, sets of china, and Limoges pottery. There were also hundreds of portraits, for which a vogue had developed during Catherine's lifetime. Many portraits in her collection were by Jean Clouet (1480–1541) and his son François Clouet (c. 1510–1572). François Clouet drew and painted portraits of all Catherine's family and of many members of the court. After Catherine's death, a decline in the quality of French portraiture set in. By 1610, the school patronised by the late Valois court and brought to its pinnacle by François Clouet had all but died out. Beyond portraiture, little is known about the painting at Catherine de' Medici's court. In the last two decades of her life, only two painters stand out as recognisable personalities: Jean Cousin the Younger (c. 1522–c. 1594), few of whose works survive, and Antoine Caron (c.

1521–1599), who became Catherine's official painter after working at Fontainebleau under Primaticcio. Caron's vivid Mannerism, with its love of ceremonial and its preoccupation with massacres, reflects the neurotic atmosphere of the French court during the Wars of Religion.

Many of Caron's paintings, such as those of the *Triumphs of the Seasons*, are of allegorical subjects that echo the festivities for which Catherine's court was famous. His designs for the Valois Tapestries celebrate the *fêtes*, picnics, and mock battles of the "magnificent" entertainments hosted by Catherine. They depict events held at Fontainebleau in 1564; at Bayonne in 1565 for the summit meeting with the Spanish court; and at the Tuileries in 1573 for the visit of the Polish ambassadors who presented the Polish crown to Catherine's son Henry of Anjou. Biographer Leonie Frieda suggests that "Catherine, more than anyone, inaugurated the fantastic entertainments for which later French monarchs also became renowned". The musical shows in particular allowed Catherine to express her creative gifts. They were usually dedicated to the ideal of peace in the realm and based on mythological themes. To create the necessary dramas, music, and scenic effects for these events, Catherine employed the leading artists and architects of the day. Historian Frances Yates has called her "a great creative artist in festivals". Catherine gradually introduced changes to the traditional entertainments: for example, she increased the prominence of dance in the shows that climaxed each series of entertainments. A distinctive new art form, the *ballet de cour*, emerged from these creative advances. Owing to its synthesis of dance, music, verse, and setting, the production of the *Ballet Comique de la Reine* in 1581 is regarded by scholars as the first authentic ballet. Catherine de' Medici's great love among the arts was architecture. "As the daughter of the Medici", suggests French art historian Jean-Pierre Babelon, "she was driven by a passion to build and a desire to leave great achievements behind her when she died." After Henry II's death, Catherine set out to immortalise her husband's memory and to enhance the grandeur of the Valois monarchy through a series of costly building projects. These included work on châteaux at Montceaux-en-Brie, Saint-Maur-des-Fossés, and Chenonceau. Catherine built two new palaces in Paris: the Tuileries and the Hôtel de la Reine. She was closely involved in the planning and supervising of all her architectural schemes.

Catherine had emblems of her love and grief carved into the stonework of her buildings. Poets lauded her as the new Artemisia, after Artemisia II of Caria, who built the Mausoleum at Halicarnassus as a tomb for her dead husband. As the centrepiece of an ambitious new chapel, she commissioned a magnificent tomb for Henry at the basilica of Saint Denis. It was designed by Francesco Primaticcio (1504–1570), with sculpture by Germain Pilon (1528–1590). Art historian Henri Zerner has called this monument "the last and most brilliant of the royal tombs of the Renaissance". Catherine also commissioned Germain Pilon to carve the marble sculpture that contains Henry II's heart. A poem by Ronsard, engraved on its base, tells the reader not to wonder that so small a vase can hold so large a heart, since Henry's real heart resides in Catherine's breast. Although Catherine spent ruinous sums on the arts, most of her patronage left no permanent legacy. The end of the Valois dynasty so soon after her death brought a change in priorities.

Titles, Styles, Honours, and Arms

Titles and Styles

- 13 April 1519-1524: Lady Caterina de' Medici
- 1524-28 October 1533: Lady Caterina de' Medici, Countess of Auvergne
- 28 October 1533-10 August 1536: The Duchess of Orléans, Countess of Auvergne
- 10 August 1536-31 March 1547: The Dauphine of France and of Viennois, Countess of Auvergne
- 31 March 1547-10 July 1559: *Her Most Christian Majesty* The Queen of France, Countess of Auvergne
- 10 July 1559-5 January 1589: *Her Most Christian Majesty* The Queen Mother of France, Countess of Auvergne

22

Mata Hari

Mata Hari was an exotic dancer from the Netherlands who gained much fame in France as a dancer. During the First World War she was arrested on charges of espionage and executed by the French by firing squad. Evidence of her actual guilt is frequently questioned. Mata Hari was born 7 August 1876 in Leeuwarden, Netherlands. Her name at birth was Margaretha Geerruida Zelle. Until the age of 13 she lived a comfortable lifestyle, attending elite private schools paid for by her father. Her father doted much attention on his 'little princess' making her used to the attention of men. However, in 1889, her father went bankrupt and her parents soon divorced. She tried to study to be a kindergarten teacher but her godfather removed her after the Headmaster became attached to Margaretha. When she was 18, she answered a newspaper ad from a handsome Dutch officer living in Java, the East Indies. In those days it was common for Dutchman living in colonies to request wives by placing ads in papers. Margaretha moved to Java where the couple had two children. However, the marriage was not a happy one, with her husband, Rudolph Macleod, suffering from alcoholism; he would often be violent to his wife, blaming her for many of his own failings. Margaretha also had difficulty in her role as housewife. She admitted "*I was not content at home... I wanted to live like a colourful butterfly in the sun.*" During her time in the Dutch East Indies Margaretha took the opportunity to learn about native dance and the local customs. Later she also had an affair with another Dutch officer, before being persuaded to return to her husband. However, they divorced in 1902, shortly after the tragic death of their young son.

When she returned to Paris in 1903, she gained work in a circus before moving on to work as an exotic dancer. With her background in the East Indies she was able to claim that she was a Java Princess of Hindu birth, something which added to her allure and fascination. One of her biographers Pat Shipman wrote of Mata Hari *"Her languid, graceful style of moving, her dark eyes and luxurious hair, telegraphed her sexuality to any male in her presence,"* It was at this time that she took the name of Mata Hari - meaning "the eye of the day" in Java. She soon became well known for her flirtatious and sensual dancing. She helped ushered in a new era of modern dance which sough inspiration from the East, Egypt and also had no inhibitions about displaying and flaunting her body. With a tradition of Eastern dance, Mata Hari helped elevate exotic dance to a more respectable background. However, it was also criticised by others for its cheap eroticism in the garb of culture. Mata Hari had several relationships with powerful men across the continent. This included Frederick William Ernest, the German crown prince, wealthy French businessmen and high ranking French military officers.

The outbreak of the First World War placed her in a unique and difficult position. Initially she found herself in Germany with her sources of finance cut off. But, with her usual resources she was able to gain sufficient finances to restart travelling. With her Dutch nationality she was able to cross different national boundaries. This took her between Germany and France often via Britain or Spain to avoid front line. As she knew both high ranking German and Frenchmen, this inevitably placed her under suspicion as someone who could in theory transfer information about the other's war effort. On one occasion she admitted to the British she was working as a spy for the French. However, the French never confirmed or denied this. In January 1917, the French intercepted a coded message from the Germans saying they had gained much useful information from a GErman spy code named H-21. It later emerged the Germans knew there code had been broken, so the Germans may have contrived to send this message framing Mata Hari, who might really have been working for French. On flimsy evidence she was convicted of spying. 30 years after the trial, one of the prosecutors was to admit there wasn't enough evidence to 'flog a dead cat'. After a trial, she was executed by firing squad on 15 October 1917. Her execution has been the source of much speculation, but sources suggest she refused a blindfold but, well dressed, she accepted her death with stoicism. Henry Wales a British reporter who covered her execution, wrote of the execution, saying when guards came to take her to the place of execution, Mata Hari merely replied *'I am ready'* Henry Wales also wrote *"Never once had the iron will of the beautiful woman failed her."* 12 soldiers were in the firing squad. After the order to shoot, she sank to her knees, an officer then shot her in the head with a pistol to make sure she was dead. After her death her life has been the subject of many accounts, some part fact, some part fictionalised

23

J.K. Rowling

> There's always room for a story that can transport people to another place. "
>
> —*J.K. Rowling*

J.K Rowling was born in Chipping Sodury, July 31st 1965. Her childhood was generally happy, although she does remember getting teased because of her name, "Rowling" – She recalls often getting called "Rowling pin" by her less than ingenious school friends. J.K. Rowling says she never really warmed to her own name, although, she does remember having a fondness tor the name Potter from quite an early age. J.K.Rowling studied at a school in Gloucestershire, before moving to Chepstow, South Wales at the age of nine. From an early age, J.K. Rowling had an ambition to be a writer. She often tried her hand at writing, although little came from her early efforts. In her own autobiography she remembers with great fondness, when her good friend Sean became the first person to give her the confidence that one day she would be able to make a very good writer.

> *"he was also the only person who thought I was bound to be a success at it, which meant much more to me than I ever told him at the time"*

Sean was also the owner of a battered old Ford Anglia, which would later appear in one of the Harry Potter series as a flying car. After finishing school, her parents encouraged her to study French at the University of Essex. She slightly regretted choosing French, saying she would have preferred to study English. However, it was her parents wish that she study something " more useful" than English. After having spent a year in Paris, J.K.Rowling graduated from university and took various jobs in London. One of her favourite jobs was working for Amnesty International; the charity, which campaigns against human rights abuses throughout the world. Amnesty International, is one of the many charities, which J.K.Rowling has generously supported since she attained a new found wealth. It was in 1990, that J.K.Rowling first conceived of the idea about Harry Potter. As she recalls, it was on a long train journey from London to Manchester when she began forming in her mind, the characters of the series. At the forefront, was a young boy, not aware that he was a wizard. The train was delayed for over four hours, but she didn't have a pen and was too shy to ask for one nothing, so nothing was written down. But she remembers being very enthusiastic, and excited about the ideas which were filling her mind.

On arriving in Manchester, she began work on writing the book immediately, although, it would take several years to come to fruition. It was also in December of 1990 that J.K.Rowling lost her mother, who died of Multiple Sclerosis. J.K.Rowling was very close to her mother, and she felt the loss deeply. Her own loss gave an added poignancy to the death of Harry Potter's mother in her book. She says her favourite scene in the *Philosopher's Stone* is, The Mirror of Erised, where Harry sees his parents in the mirror. In 1991, J.K.Rowling left England to get a job as an English teacher in Portugal. It was here that she met her first husband, and together they had a child Jessica. However, after a couple of years, the couple split after a fierce argument; where by all accounts J.K.Rowling was thrown out of the house. So she returned to England in 1994; still trying to finish her first book. She was also working full time, and bringing up her daughter as a single parent. Eventually, she finished her first copy, and sent it off to various agents.

She found an agent, Christopher, who spent over a year trying to get a publisher. Eventually, a quite small publisher, Bloomsbury agreed to take the book on. The editor Barry Cunningham also agreed to pay her an advance of £1500. The decision to take on the book was, in large part, due to his eight year old daughters enthusiastic reception of the first chapter (However she was advised to continue teaching as writers of children's books don't tend to get very well paid.)

Within a few weeks of publication, (1996) the book sales really started to take off. The initial print run was of only 1,000 – 500 of these went to libraries. First editions are now said to be worth up to £25,000 each. She also received a grant from the Scottish arts council, which enabled her to write full time. After the books initial success in the UK, an American company Scholastic agreed to pay a remarkable £100,000 for the rights to publish in America. In 1998, Warner Bros secured the film rights for the books, giving a seven figure sum. The films have magnified the success of the books, making Harry Potter into one of the most recognisable media products. Under the close guidance of J.K.Rowling, the films have sought to stay close to the original plot; also at J.K.Rowling's request all the actors are British. On the 21st December 2006, J.K.Rowling finished her final book of the Harry Potter Series – *"Harry Potter and the Deathly Hallows"*. The book was released in July 2007, becoming one of the fastest selling books of all time. J.K.Rowling has said the book is her favourite, and it makes her both happy and sad. She has said she will continue writing but there is no chance of continuing the Harry Potter Series. She however, may release a dictionary of things related to Hogwarts and Harry Potter, that were never published in other books. J.K.Rowling currently lives in Scotland, on the banks of the river Tay, with her 2nd husband Neil Murray; J.K.Rowling has 3 children, two with husband Neil.

24

Queen Elizabeth I

Queen Elizabeth I was an influential Queen of England reigning during a time of economic, political and religious upheaval. She presided over an era of economic and political expansion, which lay the framework for Britain's later dominance as a world power. It was Queen Elizabeth who also established the supremacy of Protestantism in England.

Major Achievements of Queen Elizabeth I

- United the country in a period of suspicion between Catholics and Protestants.
- Inspired troops to defeat the Spanish Armada
- Presided over a period of cultural and literary development in England.

Life of Queen Elizabeth I

Elizabeth was born in Greenwich, England on 7th September 1533. She was the daughter of Henry VIII and Anne Boleyn. Anne Boleyn was Henry's second wife. He divorced his first wife Catherine of Aragon after she had failed to produce a male heir. Unfortunately Anne Boleyn also failed to produce a male heir and would be executed for treason when Elizabeth was only 2 years old. Elizabeth was brought up at Hatfield house, Hertfordshire. Later she would be brought up in London with Catherine Parr (Henry's sixth wife) acting as step mother. As a child Elizabeth proved to be precocious and quick to learn. She excelled in academic studies and also sports; she learnt the art of public speaking, which proved to be most significant later in her reign. Following the death of Henry VIII and his only son Edward, there was uncertainty about who would inherit the throne. For 9 days a cousin of Edward, Lady Jane Grey was made queen before being disposed and then executed by Mary I. Mary's reign was unpopular as she sought to revert England to Catholicism.

Her popularity was further weakened by her distant marriage to Phillip of Spain. At one time Elizabeth's life was in danger and Mary I had her half sister arrested and kept in the Tower of London. However Elizabeth was able to convince Mary she posed no threat to her throne and eventually Mary came to trust the protestant Elizabeth and named her successor to the throne. In 1558 Mary died leaving Elizabeth as queen. Despite Mary exhorting her to retain the Catholic faith, Elizabeth ignored her wish, and she re-established Protestantism as the faith of England. However Elizabeth wished to avoid the religious extremes of Mary and Edward's reign and she sought to allow people to practise their religion of choice in private. However, later in her reign, it was alleged Catholic plotters were seeking to kill the Queen. As a consequence laws against Catholics were tightened. One figure head for the potential Catholic rebellion was Mary Queen of Scots. As a sign of her real perceived threat, Elizabeth eventually agreed to her capture and later execution (in 1587.) As a consequence of Mary's execution Catholic opposition to England grew. In particular Phillip II of Spain was determined to return Catholicism to England. There was a real threat of a Spanish invasion and in September, 1588 the powerful Spanish Armada set sail for England; threatening to make invasion a reality. Threatened with potential invasion Queen Elizabeth showed her real strength as a leader.

Her speech was enthusiastically greeted by her troops. The subsequent defeat of the heavily fortified Spanish Armada was greeted as a triumph for England and in particular Queen Elizabeth. Her personal popularity reached an all time high. It is said she was an early skilled operator of PR. She often met her subjects in person; by being highly visible she made the monarchy accessible and popular as never before. Towards the end of her reign she is reported to have said.

> *"This I account the glory of my crown, that I have reigned with your loves".*

She had many important skills as both Queen and statesman. She was quick witted, intelligent and articulate. She surrounded herself with skilled advisors and defused many potential crises. However she was also criticised for being at times both ruthless and indecisive. Several political opponents were executed for treason, although in comparison to her grandfather Henry VIII her reign was comparatively enlightened. Throughout her life she remained unmarried, despite the frequent attempts of parliament to persuade her to provide a heir. However despite many relationships with members of the court Elizabeth never gave any indication she wished to marry. For this reason she was often referred to as the "virgin queen". However her lack of direct heir meant she was the last of the Tudor monarchs. After her death the Crown passed to James I

25

Ingrid Bergman

Ingrid Bergman (29 August 1915 – 29 August 1982) was a Swedish actress who starred in a variety of European and American films. She won three Academy Awards, two Emmy Awards, and the Tony Award for Best Actress. She is ranked as the fourth greatest female star of American cinema of all time by the American Film Institute. She is best remembered for her roles as Ilsa Lund in *Casablanca* (1942), a World War II drama co-starring Humphrey Bogart and as Alicia Huberman in *Notorious* (1946), an Alfred Hitchcock thriller co-starring Cary Grant. Before becoming a star in American films, she had already been a leading actress in Swedish films. Her first introduction to American audiences came with her starring role in the English remake of *Intermezzo* in 1939. In America, she brought to the screen a "Nordic freshness and vitality", along with exceptional beauty and intelligence, and according to the *St. James Encyclopedia of Popular Culture*, she quickly became "the ideal of American womanhood" and one of Hollywood's greatest leading actresses.

Her producer David O. Selznick, who called her "the most completely conscientious actress" he had ever worked with, gave her a seven-year acting contract, thereby supporting her continued success. A few of her other starring roles, besides *Casablanca*, included *For Whom the Bell Tolls* (1943), *Gaslight* (1944), *The Bells of St. Mary's* (1945), Alfred Hitchcock's *Spellbound* (1945), *Notorious* (1946), and *Under Capricorn* (1949), and the independent production, *Joan of Arc* (1948). In 1950, after a decade of stardom in American films, she starred in the Italian film *Stromboli*, which led to a love affair with director Roberto Rossellini while they were both already married. The affair created a scandal that forced her to return to Europe until 1956, when she made a successful Hollywood comeback in *Anastasia*, for which she won her second Academy Award, as well as the forgiveness of her fans. Many of her personal and film documents can be seen in the Wesleyan University Cinema Archives.

Early Years: 1915–1938

Bergman, named after Princess Ingrid of Sweden, was born in Stockholm, Sweden on 29 August 1915 to a Swedish father, Justus Samuel Bergman, and a German mother, Friedel Adler Bergman. When she was three years of age, her mother died. Her father, who was an artist and photographer, died when she was thirteen. In the years before he died, he wanted her to become an opera star, and had her take voice lessons for three years. But she always "knew from the beginning that she wanted to be an actress", sometimes wearing her mother's clothes and staging plays in her father's empty studio. Her father documented all her birthdays with a borrowed camera. After his death, she was then sent to live with an aunt, who died of heart complications only six months later. She then moved in with her aunt Hulda and uncle Otto, who had five children. Another aunt she visited, Elsa Adler, first told Ingrid, when she was 11, that her mother may have "some Jewish blood", and that her father was aware of that fact long before they married. But her aunt also cautioned her about telling others about her Jewishness as "there might be some difficult times coming." At the age of 17, Bergman was allowed only one chance to become an actress by entering an acting competition with the Royal Dramatic Theatre in Stockholm.

Hollywood Period: 1939–1949

Bergman's first acting role in America came when Hollywood producer David O. Selznick brought her to America to star in *Intermezzo: A Love Story*, an English language remake of her 1936 Swedish film, *Intermezzo*. Unable to speak English and uncertain about her acceptance by the American audience, she expected to complete this one film and return home to Sweden. Her husband, Petter, remained in Sweden with their daughter Pia. In the film she played the role of a young piano accompanist opposite Leslie Howard as a famous violin virtuoso. She arrived in Los Angeles on 6 May 1939, and stayed at the Selznick home until she could find another residence. According to David Selznick's son Danny, who was a child at the time, his father had a few concerns about Ingrid: "She didn't speak English, she was too tall, her name sounded too German, and her eyebrows were too thick." However, Bergman was soon accepted without having to modify her looks or name, despite some early suggestions by Selznick. "He let her have her way", notes a story in *Life Magazine*. Selznick understood her fear of Hollywood make-up artists, who might turn her into someone she wouldn't recognize, and "instructed them to lay off." He was also aware that her natural good looks would compete successfully with Hollywood's "synthetic razzle-dazzle." During the weeks following, while *Intermezzo* was being filmed, Selznick was also filming *Gone with the Wind*.

Intermezzo became an enormous success and as a result Bergman became a star. The film's director, Gregory Ratoff, said "She is sensational", as an actress. This was the "sentiment of the entire set", writes *Life*, adding that workmen would go out of their way to do things for her, and the cast and crew "admired the quick, alert concentration she gave to direction and to her lines." Film historian David Thomson notes that this would become "the start of an astonishing impact on Hollywood and America" where her lack of makeup contributed to an "air of nobility." According to *Life*, the impression that she left on Hollywood, after she returned to Sweden, was of a tall (5 ft. 9 in.) girl "with light brown hair and blue eyes who was painfully shy but friendly, with a warm, straight, quick smile." Selznick appreciated her uniqueness, and with his wife Irene, they remained important friends throughout her career.

Personal Life

In 1937, at the age of 21, Bergman married dentist Petter Lindström, and a year and a half later had a daughter, Pia Lindström. After returning to America in 1940, she acted on Broadway before continuing to do films in Hollywood. The following year, her husband arrived from Sweden with daughter

Pia. Lindström stayed in Rochester, New York, where he studied medicine and surgery at the University of Rochester. Bergman would travel to New York and stay at their small rented stucco house between films, her visits lasting from a few days to four months. Dr. Lindström later moved to San Francisco, California, where he completed his internship at a private hospital, and they continued to spend time together when she could travel between filmings.

Italian Period with Rossellini: 1949–1957

Bergman strongly admired two films by Italian director Roberto Rossellini that she had seen in the United States. In 1949, Bergman wrote to Rossellini, expressing this admiration and suggesting that she make a film with him. This led to her being cast in his film *Stromboli* (1950). During production, Bergman fell in love with Rossellini, and they began an affair. Bergman became pregnant with Rossellini's son, Renato Roberto Giusto Giuseppe ("Robin") Rossellini (born 2 February 1950). This affair caused a huge scandal in the United States, where it led to Bergman being denounced on the floor of the United States Senate. Ed Sullivan chose not to have her on his show, despite a poll indicating that the public wanted her to appear. However, Steve Allen, whose show was equally popular, did have her on, later explaining "the danger of trying to judge artistic activity through the prism of one's personal life." Spoto notes that Bergman had, by virtue of her roles and screen persona, placed herself "above all that". She had played a nun in *The Bells of St. Mary's* (1945) and a saint in *Joan of Arc* (1948), and Bergman herself later acknowledged, "People saw me in *Joan of Arc* and declared me a saint. I'm not. I'm just a woman, another human being." As a result of the scandal, Bergman returned to Italy, abandoning her husband and daughter (Pia), which led to a publicized divorce and custody battle for their daughter. Bergman and Rossellini were married on 24 May 1950. In addition to Renato, they had twin daughters (born 18 June 1952): Isabella Rossellini, who became an actress and model, and Isotta Ingrid Rossellini, who became a professor of Italian literature.

Stromboli and "Neorealism"

Rossellini completed five films starring Bergman between

1949 and 1955: *Stromboli, Europa '51, Viaggio in Italia, Giovanna d'Arco al rogo*, and *La paura*. He also directed her in a brief segment of his 1953 documentary film, *Siamo donne* (We, the Women), which was devoted to film actresses. Rossellini biographer Peter Bondanella notes that problems with communication during their marriage may have inspired his films' central themes of "solitude, grace, and spirituality in a world without moral values." In addition, Rossellini's use of a Hollywood star in his typically "neorealist" films, in which he normally used non-professional actors, did provoke negative reactions in some circles.

Later Years: 1957–1982

Anastasia (1956)

With her starring role in 1956's *Anastasia*, Bergman made a triumphant return to the American screen and won the Academy Award for Best Actress for a second time. The award was accepted for her by her friend Cary Grant. Bergman would not make her first post-scandal public appearance in Hollywood until the 1958 Academy Awards, when she was the presenter of the Academy Award for Best Picture. Furthermore, after being introduced by Cary Grant and walking out on stage to present, she was given a standing ovation.

In 1972, U.S. Senator Charles H. Percy entered an apology into the *Congressional Record* for the attack made on Bergman 22 years earlier by Edwin C. Johnson. She was the President of the Jury at the 1973 Cannes Film Festival.

Murder on the Orient Express (1974)

Bergman became one of the elite actresses to receive three Oscars when she won her third (and first for Best Supporting Actress) for her performance in *Murder on the Orient Express* (1974). Director Sidney Lumet offered Bergman the important part of Princess Dragomiroff, with which he felt she could win an Oscar. However, she insisted on playing the much smaller role of Greta Ohlsson, the old Swedish missionary. Bergman could speak Swedish (her native language), German (her second language, learned from her German mother and in school), English (learned when brought over to United States), Italian (learned while living in Italy) and French (her third language, learned in school). In addition, she acted in each of these languages at various times. Fellow actor John Gielgud, who had acted with her in *Murder on the Orient Express* and who had directed her in the play *The Constant Wife*, playfully mocked this ability when he remarked, "She speaks five languages and can't act in any of them." Although known chiefly as a film star, Bergman strongly admired the great English stage actors and their craft. She had the opportunity to appear in London's West End, working with such stage stars as Sir Michael Redgrave in *A Month in the Country* (1965), Sir John Gielgud in *The Constant Wife* (1973) and Dame Wendy Hiller in *Waters of the Moon* (1977–78).

Autumn Sonata (1978)

In 1978, Bergman played in Ingmar Bergman's *Autumn Sonata* (*Höstsonaten*) for which she received her 7th Academy Award nomination and made her final performance on the big screen. In the film, Bergman plays a celebrity pianist who travels to Trondheim in Norway to visit her neglected daughter, played by Liv Ullmann.

A Woman Called Golda (1982) – Her Final Role

In 1982 she was offered the starring role in a television mini-series, *A Woman Called Golda*, about the late Israeli prime minister Golda Meir. It was to be her final acting role and she was honored posthumously with a second Emmy Award for Best Actress.

Death

Bergman died in 1982 on her 67th birthday in London, England, following a long battle with breast cancer. Her body was cremated at Kensal Green Cemetery, London and her ashes taken to Sweden, where some were scattered in the sea and the rest placed next to her parents in Norra begravningsplatsen (Northern Cemetery), Stockholm, Sweden.

26

Catherine The Great

> "Be gentle, humane, accessible, compassionate and open-handed; don't let your grandeur prevent you from mixing kindly with the humble and putting yourself in their shoes... I swear by Providence to stamp these words in my heart."
>
> —*Quote by Catherine The Great, written to herself on becoming empress in 1762*

Catherine II, called Catherine the Great (Yekaterina II Velikaya; 2 May [O.S. 21 April] 1729 – reigned as Empress of Russia from 9 July [O.S. 28 June] 1762 until 17 November [O.S. 6 November] 1796). She took power after a conspiracy deposed her husband, Peter III (1728–1762), and her reign saw the high point of the Russian nobility. Peter III, under pressure from the nobility, had already augmented the authority of the great landed proprietors over their muzhiks and serfs. In spite of the duties imposed on the nobles by the first "modernizer" of Russia, Tsar Peter I (1672–1725), and despite Catherine's friendships with the western European thinkers of the Enlightenment, Catherine found it impractical to improve the lot of her poorest subjects, who continued to suffer (for example) military conscription. The distinctions between peasant rights on votchina and pomestie estates virtually disappeared in law as well as in practice during her reign. In 1785 Catherine conferred on the nobility the Charter to the Nobility, increasing further the power of the landed oligarchs. Nobles in each district elected a Marshal of the Nobility who spoke on their behalf to the monarch on issues of concern to them — mainly economic ones.

Commentary on Life of Catherine the Great

Catherine made an unlikely Russian heroine. She was

not even Russian, but German and was thrown into a loveless marriage at the age of 14. She was tremendously disappointed in her puny husband - the rather cowardly and timid Grand Duke Peter. Catherine was a tremendously adept at wooing the great and good of the Russian court. Whilst her husband did little to attract others. Catherine threw herself into Russian culture and soon became a great socialiser, wooing friends, lovers and influencing many around her within the Russian court.. She certainly loved the power and prestige of the court, but, at the same time she had a natural sympathy with her subjects. During her time, she significantly aided the poor peasants of Russia offering many landmark social and legal improvements. She prided herself on her education, intelligence and tolerance and was an enlightened ruler for the time of history. One of her most significant lovers was Prince Potemkin, a dashing and brilliantly talent one eyed general who was also adept at politics. They may have married in secret and had a passionate love affair which lasted for many years. Together they ruled together and this led to a period of great success for Russia. Stalin later remarked that the great success of Catherine the Great was to appoint people as skilled as Prince Potemkin.

27

Mary Wollstonecraft

Mary Wollstonecraft (1759-1797) British philosopher and feminist. Best known for her book - *A Vindication of the Rights of Women* (1792) which was one of the earliest expositions of the equality of women and men. Wollstonecraft was born in 27 April 1759 in Spitalfields, London. She grew up in a difficult family situation. Her father was often violent and prone to drunken moods, especially after losing money in ill-judged investments. Mary spent much time looking after her sisters and mothers. However, in 1778, she tired of domestic life and decided to take a job as a lady's companion to Sarah Dawson. This proved a difficult experience as she didn't get on with the old lady. However, around this time, she became acquainted with Fanny Blood, who played an important role in widening Mary's horizons and ideas. The two became very close and Fanny Blood's untimely death in 1785 was quite a shock to Mary. For a while, Mary worked as a governess in a large Irish family. She had a talent for teaching, but took a dislike to Lady Kingsborough. To Mary, Kingsborough was the antithesis of an ideal women. In Lady Kingsborough she saw a women with no real independence, but being primarily concerned with superficial appearances and pleasing men. Mary later developed her thoughts for the concept of a good wife.

> "To be a good mother — a woman must have sense, and that independence of mind which few women possess who are taught to depend entirely on their husbands. Meek wives are, in general, foolish mothers; wanting their children to love them best, and take their part, in secret, against the father, who is held up as a scarecrow."
>
> —*Mary Wollstonecraft, A Vindication of the Rights of Woman (1792)*

This experience formented a desire to become a writer; Mary returned to London where she became acquainted with lumineries such as Thomas Paine, William Godwin and Joseph Johnson. In London, she became more aware of new strains in political and philosophical ideas; the late eighteenth century was an era of change. The old 'divine rights' of kings was being replaced with greater faith in human reason and liberty; this sea change in attitudes best exemplified by the French revolution. Like many radicals, Mary was initially enthused by the French revolution. In 1790, she wrote an influential pamphlet *Vindication of the Rights of Men* (1790). This sought to defend the principles of the French Revolution against Edmund Burke's conservative critique. This helped establish Mary as a leading liberal writer; at the time, it was rare for a women to have such prominence in literary circles.

Vindication of the Rights of Women

Shortly after the *Vindication of the Rights of Men* (1970), Mary wrote *A Vindication of the Rights of Women* (1972). This was groundbreaking work, as it proposed women were the equal of men. Wollstonecraft contended, it was only the lack of education for women that meant they seemed to be intellectually inferior.

> "Till women are more rationally educated, the progress in human virtue and improvement in knowledge must receive continual checks."

A Vindication of the Rights of Woman (1792) Ch 3. She was highly critical of the contemporary attitudes to women

> "Women are systematically degraded by receiving the trivial attentions which men think it manly to pay to the sex, when, in fact, men are insultingly supporting their own superiority."

A Vindication of the Rights of Woman (1792) Ch 3.

Such arguments were radical for the time. Even liberal authors didn't wholeheartedly agree with its arguments and beliefs. After its publication, Mary visited revolutionary Paris. However, the situation quickly deteriorated, Louis XVI was guillotined and the revolution became increasingly repressive. In Paris she fell madly in love with an American, Gilbert Imlay. Together they had an illegitimate daughter. When

Britain and France declared war on each other, Mary needed the protection of appearing to be married to an American to prevent her being arrested. Together they had a child, Fanny. However, the relationship grew increasingly difficult, as Gilbert proved to fall well short of Mary's romantic ideal. Overcome with grief, they broke up and Mary returned to England. On returning to London, she attempted to commit suicide in despair at the failed relationship (she also experienced depressive moods throughout her life). However, the attempt failed, and she was rescued from the River Thames by a passing man.

Mary Wollstonecraft and William Godwin

After slowly recovering from the depths of depression, Mary restarted her literary career and become romantically involved with William Godwin. Mary became pregnant and the two decided to get married. Tragically, Mary died in childbirth, though her daughter (Mary Godwin) survived and went on to become the author of *Frankenstein* and wife of Percy Shelley. After her death, William Godwin published her memoirs which proved quite shocking to society. People were not comfortable with the unorthodox and free-living attitude of Mary Wollstonecraft. Even in the late Victorian suffragette movement, Mary Wollstonecraft was given a low profile, as her life sat uncomfortably with Victorian attitudes. However, by the Twentieth Century, Wollstonecraft's writings were seen as key developments in the concept of women's rights. In many ways, Mary Wollstonecraft was many years, if not centuries ahead of her time.

28

Jane Austen

Jane Austen was the author of several enduringly popular English novels, including: Pride and Prejudice, Emma and Mansfield Park.

Early Life

Jane Austen was born in Steventon, Hampshire on 16th December 1775. She was the 7th daughter of an 8 child family. Her father, George Austen, was a vicar and lived on a reasonable income of £600 a year. However, although they were middle class, they were not rich; her father would have been unable to give much to help her daughters get married. Jane was brought up with her 5 brothers and her elder sister Cassandra. (another brother, Edward, was adopted by a rich, childless couple and went to live with them). Jane was close to her siblings, especially Cassandra, to whom she was devoted. The two sisters shared a long correspondence throughout her life; much of what we know about Jane comes from these letters, although, unfortunately Cassandra burnt a number of these on Jane's death. Jane was educated at Oxford and later a boarding school in Reading. In the early 1800s two of Jane's brother's joined the navy, leaving to fight in the Napoleonic wars; they would go on to become admirals. The naval connections can be seen in novels like Mansfield Park. After the death of her father in 1805, Jane, with her mother and sister returned to Hampshire. In 1809, her brother, Edward who had been brought up by the Knights, invited the family to the estate he had inherited at Chawton. It was in the country house of Chawton, that Jane was able to produce some of her greatest novels.

Novels of Jane Austen

Her novels are a reflection of her outlook on life. She spent most of her life insulated from certain sections of society. Her close friends were mainly her family, and those of similar social standing. It is not surprising then that her novels focused on 2 or 3 families of the middle or upper classes. Most novels were also based on the idyll of rural country houses that Jane was so fond of. Her novels also focus on the issue of gaining a suitable marriage. Marriage was a big issue facing women and men of her time; often financial considerations were paramount in deciding marriages. As an author, Jane used to satirise these financial motivations, for example, in Pride and Prejudice the mother is ridiculed for her ambitions to marry her daughters for maximum financial remuneration. Jane, herself remained single throughout her life. Apart from brief flirtations, Jane remained single, and appeared to have little interest in getting married (unlike the characters of her novels. The strength of Jane's novels was her ability to gain penetrating insights into the character and nature of human relationships, from even a fairly limited range of environments and characters.

King George IV actually requested that one novel could be dedicated to him. Emma is therefore dedicated to the King, even though Jane did not maintain any liking towards the King. Not all were favourable to Jane.

Death of Jane Austen

Jane died in 1816, aged only 41. She died of Addison's disease, a disorder of the adrenal glands. She was buried at Winchester Cathedral.

There are two museums dedicated to Jane Austen.

- The Jane Austen Centre in Bath and
- The Jane Austen's House Museum, located in Chawton cottage, in Hampshire, where she lived from 1809 – 1816

In 2005, Pride and Prejudice was voted best British novel of all time in a BBC poll.

29

Aung San Suu Kyi

Aung San Suu Kyi (born 19 June 1945) is a Burmese opposition politician and a former General Secretary of the National League for Democracy. In the 1990 general election, Aung San Suu Kyi's National League for Democracy party won 59% of the national votes and 81% (392 of 485) of the seats in Parliament. She had, however, already been detained under house arrest before the elections. She remained under house arrest in Burma for almost 15 of the 21 years from 20 July 1989 until her release on 13 November 2010. Primarily in response to her detention, Aung San Suu Kyi received the Rafto Prize and the Sakharov Prize for Freedom of Thought in 1990 and the Nobel Peace Prize in 1991. In 1992 she was awarded the Jawaharlal Nehru Award for International Understanding by the government of India and the International Simón Bolívar Prize from the government of Venezuela. In 2007, the Government of Canada made her an honorary citizen of that country, one of only five people ever to receive the honor. Aung San Suu Kyi is the third child and only daughter of Aung San, considered to be the father of modern-day Burma.

Aung San Suu Kyi derives her name from three relatives: "Aung San" from her father, "Suu" from her paternal grandmother and "Kyi" from her mother Khin Kyi. She is frequently called Daw Aung San Suu Kyi. *Daw* is not part of her name, but is an honorific, similar to madame, for older, revered women, literally meaning "aunt". She is also often referred to as Daw Suu by the Burmese, or "Aunty Suu", and as Dr. Suu Kyi, Ms. Suu Kyi, or Mrs. Suu Kyi by the foreign media. However, like other Burmese, she has no surname. Her name is pronounced Awn Sahn Sue Chee.

Personal Life

Aung San Suu Kyi was born in Rangoon (now named Yangon). Her father, Aung San, founded the modern Burmese army and negotiated Burma's independence from the British Empire in 1947; he was assassinated by his rivals in the same year. She grew up with her mother, Khin Kyi, and two brothers, Aung San Lin and Aung San Oo, in Rangoon. Aung San Lin, died at age eight, when he drowned in an ornamental lake on the grounds of the house. Her elder brother emigrated to San Diego, California, becoming a United States citizen. After Aung San Lin's death, the family moved to a house by Inya Lake where Suu Kyi met people of very different backgrounds, political views and religions. She was educated in Methodist English High School (now Basic Education High School No. 1 Dagon) for much of her childhood in Burma, where she was noted as having a talent for learning languages. She is a Theravada Buddhist. Suu Kyi's mother, Khin Kyi, gained prominence as a political figure in the newly formed Burmese government. She was appointed Burmese ambassador to India and Nepal in 1960, and Aung San Suu Kyi followed her there, graduating from Lady Shri Ram College in New Delhi with a degree in politics in 1964. Suu Kyi continued her education at St Hugh's College, Oxford, obtaining a B.A. degree in Philosophy, Politics and Economics in 1969. After graduating, she lived in New York City with a family friend and worked at the UN for three years, primarily on budget matters, writing daily to her future husband, Dr. Michael Aris. In 1972, Aung San Suu Kyi married Aris, a scholar of Tibetan culture, living abroad in Bhutan.

The following year she gave birth to their first son, Alexander Aris, in London; their second son, Kim, was born in 1977. Subsequently, she earned a PhD at the School of Oriental and African Studies, University of London in 1985. She was elected an Honorary Fellow in 1990. For two years she was a Fellow at the Indian Institute of Advanced Studies (IIAS) in Shimla, India. She also worked for the government of the Union of Burma. In 1988 Suu Kyi returned to Burma, at first to tend for her ailing mother but later to lead the pro-democracy movement. Aris' visit in Christmas 1995 turned out to be the last time that he and Suu Kyi met, as Suu Kyi remained in Burma and the Burmese dictatorship denied him any further entry visas. Aris was diagnosed with prostate cancer in 1997 which was later found to be terminal. Despite appeals from prominent figures and organizations, including the United States, UN Secretary General Kofi Annan and Pope John Paul II, the Burmese government would not grant Aris a visa, saying that they did not have the facilities to care for him, and instead urged Aung San Suu Kyi to leave the country to visit him. She was at that time temporarily free from house arrest but was unwilling to depart, fearing that she would be refused re-entry if she left, as she did not trust the military junta's assurance that she could return. Aris died on his 53rd birthday on 27 March 1999. Since 1989, when his wife was first placed under house arrest, he had seen her only five times, the last of which was for Christmas in 1995. She also remains separated from her children, who live in the United Kingdom. On 2 May 2008, after Cyclone Nargis hit Burma, Suu Kyi lost the roof of her house and lived in virtual darkness after losing electricity in her dilapidated lakeside residence. She used candles at night as she was not provided any generator set. Plans to renovate and repair the house were announced in August 2009. Suu Kyi was released from house arrest on 13 November 2010.

Political Beginning and Career

Aung San Suu Kyi returned to Burma in 1988 to take care of her ailing mother. By coincidence, in the same year, the long-time military leader of Burma and head of the ruling party, General Ne Win, stepped down. This led to mass demonstrations for democracy on 8 August 1988 (8–8–88, a day seen as auspicious), which were violently suppressed in what came to be known as the 8888 Uprising. On 26 August 1988, she addressed half a million people at a mass rally in front of the Shwedagon Pagoda in the capital, calling for a democratic government. However in September, a new military junta took power. Later the same month, 24 September 1988, the National League for Democracy (NLD) was formed, with Suu Kyi as general secretary. Influenced by both Mahatma Gandhi's philosophy of non-violence and by more specifically Buddhist concepts, Aung San Suu Kyi entered politics to work for democratization, helped found the National League for Democracy on 24 September 1988, and was put under house arrest on 20 July 1989.

International Support

Aung San Suu Kyi has received vocal support from Western nations in Europe, Australia and North and South America, as well as India, Japan and South Korea. In December 2007, the US House of Representatives voted unanimously 400–0 to award Aung San Suu Kyi the Congressional Gold Medal; the Senate concurred on 25 April 2008. On 6 May 2008, President George Bush signed legislation awarding Suu Kyi the Congressional Gold Medal. She is the first recipient in American history to receive the prize while imprisoned. More recently, there has been growing criticism of her detention by Burma's neighbours in the Association of Southeast Asian Nations, particularly from Indonesia, Thailand, the Philippines and Singapore. At one point Malaysia warned Burma faced expulsion from ASEAN as a result of the detention of Suu Kyi. Other nations including South Africa, Bangladesh and the Maldives have also called for her release.

However, Samak Sundaravej, former Prime Minister of Thailand, criticised the amount of support for Suu Kyi, saying that "Europe uses Aung San Suu Kyi as a tool. If it's not related to Aung San Suu Kyi, you can have deeper discussions with Myanmar." U2 supported her on their 2009 U2 360° Tour by encouraging fans to wear masks with her likeness on them during the band's performance of the song "Walk On", which was originally written for her, and in some cities they welcomed Amnesty International volunteers on stage again during the performance of the song, carrying lanterns in her tribute. In 2005, Irish singer songwriter Damien Rice released the single Unplayed Piano in support of Aung San Suu Kyi. Vietnam, however, does not support calls by other ASEAN member states for Myanmar to free Aung San Suu Kyi, state media reported Friday, 14 August. 2009. The state-run ViÇt Nam News said Vietnam had no criticism of Myanmar's decision 11 August 2009 to place Suu Kyi under house arrest for the next 18 months, effectively barring her from elections scheduled for 2010. "It is our view that the Aung San Suu Kyi trial is an internal affair of Myanmar", Vietnamese government spokesman Le Dung stated on the website of the Ministry of Foreign Affairs.

Nobel Peace Prize

Aung San Suu Kyi was awarded the Nobel Peace Prize in 1991. The decision of the Nobel Committee mentions: The Norwegian Nobel Committee has decided to award the Nobel Peace Prize for 1991 to Aung San Suu Kyi of Myanmar (Burma) for her non-violent struggle for democracy and human rights. Nobel Peace Prize winners (Archbishop Desmond Tutu, The Dalai Lama, Shirin Ebadi, Adolfo Pérez Esquivel, Mairead Corrigan, Rigoberta Menchú, Prof. Elie Wiesel, U.S. President Barack Obama, Betty Williams, Jody Williams and former U.S. President Jimmy Carter) called for the rulers of Burma to release Suu Kyi "create the necessary conditions for a genuine dialogue with Daw Aung San Suu Kyi and all concerned parties and ethnic groups in order to achieve an inclusive national reconciliation with the direct support of the United Nations."

Books

Authored

- *Der Weg zur Freiheit* (1999) with U Kyi Maung, U Tin Oo, ISBN 978-3404614356
- *Letters from Burma* (1998) with Fergal Keane ISBN 978-0140264036
- *The Voice of Hope* (1998) with Alan Clements, ISBN 978-1888363838, fully updated and re-issued in October 2008 by Rider Books, ISBN 978-1846041433
- *Letter to Daniel: Despatches from the Heart* (1996) by Fergal Keane, foreword by Aung San Suu Kyi, edited by Tony Grant ISBN 978-0140262896
- *Freedom from Fear and other Writings* (1995) with Václav Havel, Desmond M. Tutu, and Michael Aris, ISBN 978-0140253177
- *Burma's Revolution of the Spirit: The Struggle for Democratic Freedom and Dignity* (1994) with Alan Clements, Leslie Kean, the Dalai Lama, Sein Win, ISBN 978-0893815806
- *Aung San of Burma: A Biographical Portrait by His*

Daughter (1991) ISBN 978-1870838801, 2nd edition 1995

Awards

- Thorolf Rafto Memorial Prize (1990)
- Sakharov Prize (1990)
- Nobel Peace Prize (1991)
- Simón Bolívar International Prize (1992)
- Jawaharlal Nehru Award for International Understanding (1993)
- Prize For Freedom of the Liberal International (1995)
- Honorary Companion of the Order of Australia (1996)
- Doctor of Laws Honorary degree from the University of Bath (1998)
- Freedom of Dublin City,Ireland (1999)
- Presidential Medal of Freedom (2000)
- UNESCO Madanjeet Singh Prize for the Promotion of Tolerance & Non-Violence (2002)
- Gwangju Prize for Human Rights (2004)
- Olof Palme Prize (2005)
- Doctor of Laws (honoris causa) from Memorial University of Newfoundland (2004)
- Freedom from Fear award (2006)
- Honorary Canadian citizenship (2007)
- Honorary President of the LSESU (2007)
- Doctorate of Letters (honoris causa) from Colgate University (2008)
- Congressional Gold Medal (2008)
- Premi Internacional Catalunya (2008)
- Honorary Member of the Club of Madrid (2008)
- Freedom Of Glasgow (2009)
- Mahatma Gandhi International Award for Peace and Reconciliation (2009)
- Honorary Doctor of Laws from University of Ulster in recognition of her services to human rights (2009)
- Ambassador of Conscience Award (2009) from Amnesty International

30

Harriet Beecher Stowe

Harriet Beecher Stowe (June 14, 1811 – July 1, 1896) was an American abolitionist and author. Her novel *Uncle Tom's Cabin* (1852) depicted life for African-Americans under slavery; it reached millions as a novel and play, and became influential in the United States and United Kingdom. It energized anti-slavery forces in the American North, while provoking widespread anger in the South. She wrote more than 20 books, including novels, three travel memoirs, and collections of articles and letters. She was influential both for her writings and her public stands on social issues of the day.

Biography

Harriet Elisabeth Beecher was born in Litchfield, Connecticut on June 14, 1811. She was the middle daughter of three, born to outspoken religious leader Lyman Beecher and Roxana Foote, a deeply religious woman who died when Stowe was only five years old. Her older sister was the educator and author, Catharine Beecher, and her younger sister was Isabella, who married the attorney John Hooker and had a family. They had seven brothers, all of whom became ministers: including Henry Ward Beecher, Charles Beecher, and Edward Beecher. Harriet enrolled in the seminary (girls' school) run by her sister Catharine, where she received a traditionally "male" education in the classics, including study of languages and mathematics. Among her classmates there was Sarah P. Willis, who later wrote under the pseudonym Fanny Fern. At the age of 21, she moved to Cincinnati, Ohio to join her father, who had become the president of Lane Theological Seminary. There, she also joined the Semi-Colon Club, a literary salon and social club whose members included the Beecher sisters,

Caroline Lee Hentz, Salmon P. Chase, Emily Blackwell, and others. It was in that group that she met Calvin Ellis Stowe, a widower and professor at the seminary. The two married on January 6, 1836. He was an ardent critic of slavery, and the Stowes supported the Underground Railroad, temporarily housing several fugitive slaves in their home. They had seven children together, including twin daughters.

Later Years

In the 1870s, Stowe's brother Henry Ward Beecher was accused of adultery and became the subject of a national scandal. Stowe, unable to bear the public attacks on her brother, fled to Florida but asked family members to send her newspaper reports. Through the affair, however, she remained loyal to her brother and believed he was innocent. Mrs. Stowe was among the founders of the Hartford Art School which later became part of the University of Hartford. Stowe died on July 1, 1896, at age eighty-five, in Hartford, Connecticut. She is buried in the historic cemetery at Phillips Academy in Andover, Massachusetts.

Legacy

Landmarks

The Harriet Beecher Stowe House in Cincinnati, Ohio is the former home of her father Lyman Beecher on the former campus of the Lane Seminary. Her father was a preacher who was greatly affected by the pro-slavery Cincinnati Riots of 1836. Harriet Beecher Stowe lived here until her marriage. It is open to the public and operated as a historical and cultural site, focusing on Harriet Beecher Stowe, the Lane Seminary and the Underground Railroad. The site also presents African-American history. In the 1870s and 1880s, Stowe and her family wintered in Mandarin, Florida, now a suburb of modern consolidated Jacksonville, on the St. Johns River. Stowe wrote *Palmetto Leaves* while living in Mandarin, arguably an eloquent piece of promotional literature directed at Florida's potential Northern investors at the time. The book was published in 1873 and describes Northeast Florida and its residents. In 1870, Stowe created an integrated school in Mandarin for children and adults. This predated the national movement toward integration by more than a half century. The marker commemorating the Stowe family is located across the street from the former site of their cottage. It is on the property of the Community Club, at the site of a church where Stowe's husband once served as a minister. The Harriet Beecher Stowe House in Brunswick, Maine is where Stowe wrote *Uncle Tom's Cabin*. She and her husband lived here while he worked at Bowdoin College. Although local interest has been strong to preserve the house as a museum, it has long been in private ownership and operated as an inn and German restaurant. It most recently changed ownership in 1999 for $865,000.

The Harriet Beecher Stowe House in Hartford, Connecticut is the house where Stowe lived for the last 23 years of her life. It was next door to the house of fellow author Mark Twain. In this 5,000 sq ft (460 m) cottage-style house, there are many of Beecher Stowe's original items and items from the time period. In the research library, which is open to the public, there are numerous letters and documents from the Beecher family. The house is opened to the public and offers house tours on the half hour. In 1833, during Stowe's time in Cincinnati, the city was afflicted with a serious cholera epidemic. To avoid illness, Stowe made a visit to Washington, Kentucky, a major community of the era just south of Maysville. She stayed with the Marshall Key family, one of whose daughters was a student at Lane Seminary. It is recorded that Mr. Key took her to see a slave auction, as they were frequently held in Maysville. Scholars believe she was strongly moved by the experience. The Marshall Key home still stands in Washington. Key was a prominent Kentuckian; his visitors also included Henry Clay and Daniel Webster. The Uncle Tom's Cabin Historic Site is part of the restored Dawn Settlement at Dresden, Ontario, which is 20 miles east of Algonac, Michigan. The community for freed slaves founded by the Rev. Josiah Henson and other abolitionists in the 1830s has been restored. There's also a museum. Henson and the Dawn Settlement provide Stowe with the inspiration for *Uncle Tom's Cabin*.

Honors

- Stowe is honored with a feast day on the liturgical calendar of the Episcopal Church (USA) on July 1.

- On June 13, 2007, the United States Postal Service issued a 75¢ Distinguished Americans series postage stamp in her honor.
- In early 2010, Stowe was proposed by the Ohio Historical Society as a finalist in a statewide vote for inclusion in Statuary Hall at the United States Capitol.

Partial List of Works

- *The Mayflower; or, Sketches of Scenes and Characters Among the Descendants of the Pilgrims* (1834)
- *Uncle Tom's Cabin* (1852)
- *A Key to Uncle Tom's Cabin* (1853)
- *Dred, A Tale of the Great Dismal Swamp* (1856)
- *The Minister's Wooing* (1859)
- *Agnes of Sorrento* (1862) (reading online)
- *The Pearl of Orr's Island* (1862)
- *Old Town Folks* (1869)
- *Little Pussy Willow* (1870)
- *Lady Byron Vindicated* (1870)
- *My Wife and I* (1871)
- *Pink and White Tyranny* (1871)
- *Woman in Sacred History* (1873)
- *Palmetto Leaves* (1873)
- *We and Our Neighbors* (1875)
- *Poganuc People* (1878)
- *The Poor Life* (1890)

As Christopher Crowfield

- *House and Home Papers* (1865)
- *Little Foxes* (1866)
- *The Chimney Corner* (1868)

31

Mae West

Mae West (August 17, 1893 – November 22, 1980) was an American actress, playwright, screenwriter and sex symbol. Known for her bawdy double entendres, West made a name for herself in Vaudeville and on the stage in New York before moving to Hollywood to become a comedienne, actress and writer in the motion picture industry. One of the more controversial movie stars of her day, West encountered many problems including censorship. When her cinematic career ended, she continued to perform on stage, in Las Vegas, in the United Kingdom, on radio and television, and recorded rock and roll albums. She used the alias Jane Mast early in her career.

Early Life and Career

West was born Mary Jane West in Bushwick, Brooklyn, Kings County, New York, daughter of John Patrick West and Matilda "Tillie" Doelger (also spelled Delker). Her father was a prizefighter known as "Battlin' Jack West" who later worked as a "special policeman" and then as a private investigator who ran his own agency. Her mother was a former corset and fashion model. The family was Protestant, although by some accounts West's mother was a Jewish immigrant from Bavaria. Her paternal grandmother was an Irish Catholic, but West's paternal grandfather, John Edwin, may have been an African American who passed for white. Her siblings were Mildred Katherine West (December 8, 1898 – March 12, 1982), known as Beverly, and John Edwin West (February 11, 1900 – October 12, 1964). During her childhood, West's family moved to various parts of Woodhaven, Queens, as well as Williamsburg and

Greenpoint in Brooklyn. She may have attended Erasmus Hall High School. At five years old, West first entertained a crowd, at a church social, and she started appearing in amateur shows at the age of seven. She often won prizes at local talent contests. She began performing professionally in vaudeville in the Hal Clarendon Stock Company in 1907 at the age of fourteen. West first performed under the stage name *Baby Mae*, and tried various personas including a male impersonator, Sis Hopkins, and a blackface coon shouter. Her trademark walk was said to have been inspired or influenced by female impersonators Bert Savoy and Julian Eltinge, who were famous during the Pansy Craze. Her first appearance in a legitimate Broadway show was in a 1911 revue *A La Broadway* put on by her former dancing teacher, Ned Wayburn.

The show folded after just eight performances. She then appeared in a show called *Vera Violetta*, whose cast featured Al Jolson. In 1912 she also appeared in the opening performance of "A Winsome Widow" as a 'baby vamp' named La Petite Daffy. Her photograph appeared on an edition of the sheet music for the popular number "Ev'rybody Shimmies Now" in 1918. She was encouraged as a performer by her mother, who, according to West, always thought that whatever her daughter did was fantastic. However, other members of her family were less encouraging, including the aunt who delivered her, as well as her paternal grandmother. Along with certain other relatives, they are all reported as having disapproved of her career and her choices. In 1918, after exiting several high-profile revues, West finally got her break in the Shubert Brothers revue *Sometime*, opposite Ed Wynn. Her character Mayme danced the shimmy. Eventually, she began writing her own risqué plays using the pen name Jane Mast. Her first starring role on Broadway was in a play she titled *Sex*, which she also wrote, produced, and directed. Though critics hated the show, ticket sales were good. The notorious production did not go over well with city officials and the theater was raided with West arrested along with the cast.

She was prosecuted on morals charges and, on April 19, 1927, was sentenced to ten days for "corrupting the morals of youth." While incarcerated on Welfare Island (now known as

Roosevelt Island), she dined with the warden and his wife and told reporters that she wore her silk underpants while serving time. She served eight days with two days off for good behavior. Media attention about the case enhanced her career. Her next play, *The Drag*, dealt with homosexuality and was what West called one of her "comedy-dramas of life". After a series of try-outs in Connecticut and New Jersey, West announced she would open the play in New York. However, *The Drag* never opened on Broadway due to the Society for the Prevention of Vice vows to ban it if West attempted to stage it. West was an early supporter of the women's liberation movement, but stated she was not a feminist. She was also a supporter of gay rights. West continued to write plays, including *The Wicked Age*, *Pleasure Man* and *The Constant Sinner*. Her productions were plagued by controversy and other problems, although the controversy ensured that West stayed in the news and most of the time this resulted in packed performances. Her 1928 play, *Diamond Lil*, about a racy, easygoing lady of the 1890s, became a Broadway hit. This show enjoyed an enduring popularity and West would successfully revive it many times throughout the course of her career.

Motion Pictures

In 1932, West was offered a motion picture contract by Paramount Pictures, when she was 38 years old (although she kept her age ambiguous for several more years). She made her film debut in *Night After Night* starring George Raft. At first, she did not like her small role in *Night After Night*, but was appeased when she was allowed to rewrite her scenes. In West's first scene, a hat check girl exclaims, "Goodness, what beautiful diamonds." West replies, "Goodness had nothing to do with it, dearie." Reflecting on the overall result of her rewritten scenes, Raft is said to have remarked, "She stole everything but the cameras." She brought her *Diamond Lil* character, now renamed *Lady Lou*, to the screen in *She Done Him Wrong* (1933). The film is also notable as one of Cary Grant's first major roles, which boosted his career. West claimed she spotted Grant at the studio and insisted that he be cast as the male lead. The film was a box office hit and earned an Academy Award nomination for Best Picture. The success of the film most likely saved Paramount from bankruptcy. Her next release, *I'm No Angel* (1933), paired her with Grant again. *I'm No Angel* was also a financial success. By 1933, West was the eighth-largest U.S. box office draw in the United States and, by 1935, the second-highest paid person in the United States (after William Randolph Hearst). On July 1, 1934, the censorship of the Production Code began to be seriously and meticulously enforced, and her screenplays were heavily edited.

West's next film was *Belle of the Nineties* (1934). Originally titled *It Ain't No Sin*, the title was changed due to the censors' objections. Her next film, *Goin' to Town* (1935), received mixed reviews. Her next film, *Klondike Annie* (1936) dealt, as best it could given the heavy censorship, with religion and hypocrisy. Some critics called the film her screen masterpiece. That same year, West played opposite Randolph Scott in *Go West, Young Man*. In this film, she adapted Lawrence Riley's Broadway hit *Personal Appearance* into a screenplay. Directed by Henry Hathaway, *Go West, Young Man* is considered one of West's weaker films of the era. After this film, West starred in *Every Day's a Holiday* (1937) for Paramount before their association came to an end. In 1939, Universal Pictures approached West to star in a film opposite W. C. Fields. The studio was eager to duplicate the success of *Destry Rides Again* starring Marlene Dietrich and James Stewart with a vehicle starring West and Fields. Having left Paramount eighteen months earlier and looking for a comeback film, West accepted the role of Flower Belle Lee in the film *My Little Chickadee* (1940). Despite their intense mutual dislike, and fights over the screenplay, *My Little Chickadee* was a box office success, outgrossing Fields' previous films *You Can't Cheat an Honest Man* (1939) and *The Bank Dick* (1940). West's next film was *The Heat's On* (1943) for Columbia Pictures. She initially didn't want to do the film but after producer and director Gregory Ratoff pleaded with her and claimed he would go bankrupt if she didn't, West relented. The film opened to bad reviews and failed at the box office. West would not return to films until 1970.

Radio

On December 12, 1937, West appeared in two separate

sketches on ventriloquist Edgar Bergen's radio show *The Chase and Sanborn Hour*. By this point in her career, West's fame was fading and she was on the show hoping to promote her latest film, *Every Day's a Holiday* Appearing as herself, West flirted with Charlie McCarthy, Bergen's dummy, using her usual brand of wit and risqué sexual references. West referred to Charlie as "all wood and a yard long" and commented that his kisses gave her splinters. Even more outrageous was a sketch written by Arch Oboler that starred West and Don Ameche as Adam and Eve in the Garden of Eden. She told Ameche in the show to "get me a big one... I feel like doin' a big apple!" Days after the broadcast, NBC received letters calling the show "immoral" and "obscene". Women's clubs and Catholic groups admonished the show's sponsor, Chase & Sanborn Coffee Company, for "prostituting" their services for allowing "impurity [to] invade the air". The Federal Communications Commission (FCC) later deemed the broadcast "vulgar and indecent" and "far below even the minimum standard which should control in the selection and production of broadcast programs". There is some debate regarding the reaction to the skit, however. Mainstream reaction was not as swift as that of Catholics. Some claim that Catholic groups already had it in for Mae West; they despised her sexual image and warned the sponsor of the program they were planning to protest. Nevertheless, the incident is known as one of the first cases where radio programming faced claims of indecency from the FCC. NBC personally blamed West for the incident and banned her (and the mention of her name) from their stations. They claimed it was not the content of the skit, but West's tonal inflections that gave it the controversial context. West would not perform in radio for another twelve years until January 1950, in an episode of *The Chesterfield Supper Club* hosted by Perry Como.

Middle Years

After appearing in *The Heat's On* in 1943, West remained active during the ensuing years. Among her stage performances was the title role in *Catherine was Great* (1944) on Broadway, in which she spoofed the story of Catherine the Great of Russia, surrounding herself with an "imperial guard" of tall, muscular young actors. The play was produced by Mike Todd and ran for 191 performances. In the 1950s, she also starred in her own Las Vegas stage show, singing while surrounded by bodybuilders. Jayne Mansfield met and later married one of West's muscle men, a former Mr. Universe, Mickey Hargitay. When casting the role of Norma Desmond for the 1950 film *Sunset Boulevard*, Billy Wilder offered West, then nearing 60, the role. West turned down the part. Wilder later said, "The idea of [casting] Mae West was idiotic because we only had to talk to her to find out that she thought she was as great, as desirable, as sexy as she had ever been." Gloria Swanson was eventually cast in the role. In 1958, West appeared at the Academy Awards and performed the song "Baby, It's Cold Outside" with Rock Hudson. In 1959, she released her autobiography entitled *Goodness Had Nothing to Do with It*, which went on to become a best seller.

Later Career and Final Years

West made some rare appearances on television, including *The Red Skelton Show* in 1960. In 1964, she guest starred on the sitcom *Mister Ed.* In order to keep her appeal fresh with younger generations, she recorded two rock and roll albums, *Way Out West* and *Wild Christmas* in the late 1960s. She also recorded a number of parody songs including "Santa, Come Up to See Me" on the album *Wild Christmas.* After a 26-year absence from motion pictures, West appeared as Leticia Van Allen in Gore Vidal's *Myra Breckinridge* (1970) with Raquel Welch, Rex Reed, Farrah Fawcett, and Tom Selleck in a small part. The movie was a deliberately campy sex change comedy that was both a box office and critical failure. Vidal later called the film "an awful joke". Despite *Myra Breckinridge*'s mainstream failure, it did find an audience on the cult film circuit where West's films were regularly screened and West herself was dubbed "the queen of camp". West recorded another album in the 1970s on MGM Records titled *Great Balls of Fire*, which covered songs by The Doors among others. Her autobiography, *Goodness Had Nothing to Do with It,* was also updated and republished.

In 1976, she appeared on *The Dick Cavett Show* and that same year began work on her final film, *Sextette* (1978). Adapted from a script written by West, daily revisions and disagreements hampered production from the beginning. Due to the numerous changes, West agreed to have her lines fed to her through a speaker concealed in her wig. Despite the daily problems, West was, according to *Sextette* director Ken Hughes, determined to see the film through. In spite of her determination, Hughes noted that West sometimes appeared disoriented and forgetful and found it difficult to follow his directions. Her now failing eyesight also made navigating around the set difficult. Hughes eventually began shooting her from the waist up to hide the out-of-shot production assistant crawling on the floor, guiding her around the set. Upon its release, *Sextette* was a critical and commercial failure. In August 1980, West tripped while getting out of bed. After the fall, West was unable to speak and was taken to the Good Samaritan Hospital in Los Angeles where tests revealed that she had suffered a stroke. She remained in the hospital where, seven days later, she had a diabetic reaction to the formula in her feeding tube. On September 18, she suffered a second stroke which left her right side paralyzed and developed pneumonia. By November, her condition had improved, but the prognosis was not good and she was sent home. She died there on November 22, 1980, at age 87. A private service was held in the Old North Church replica, in Forest Lawn, Hollywood Hills, on November 25, 1980. Bishop Andre Penachio, who was also a friend, officiated at the entombment in the family room at Cypress Hills Abbey, Brooklyn, purchased in 1930 when her mother died. Her father and brother were also entombed there before her, and her younger sister was laid to rest in the last of the five crypts within 18 months after West's death. For her contribution to the film industry, she has a star on the Hollywood Walk of Fame at 1560 Vine Street in Hollywood.

Personal Life

West was married on April 11, 1911, in Milwaukee, Wisconsin, to Frank Szatkus, stage name Frank Wallace, a fellow vaudevillian whom she first met in 1909. She was 17, he was 21. West kept the marriage a secret. But in 1935, after West had made several hit movies, a filing clerk discovered

West's marriage certificate and alerted the press. An affidavit in which she had declared herself married, which she made during the *Sex* trial in 1927, was also uncovered. At first, West denied ever marrying Wallace but finally admitted in July 1937, in reply to a legal interrogatory, that they had been married. Even though the marriage was a reality, she never lived with Wallace as husband and wife. She insisted they have separate bedrooms and she soon sent him away in a show of his own in order to get rid of him. She obtained a legal divorce on July 21, 1942, during which Wallace withdrew his request for separate maintenance, and West testified that she and Wallace had lived together for only "several weeks." The final divorce decree was granted on May 7, 1943. West may also have had another secret marriage. In August 1913, she met an Italian-born Vaudeville headliner and star of the piano-accordion, Guido Deiro. Her affair went "[v]ery deep, hittin' on all the emotions. You can't get too hot over anybody unless there's somethin' that goes along with the sex act, can you?" Deiro fell in love with West and arranged his bookings so that the two traveled together. They became engaged in either late 1913 or early 1914. Some sources reported the pair were married.

During a 1935 radio broadcast Walter Winchell incorrectly reported that Mae West had been married to Guido's brother, Pietro. Walter Wincher, a writer for *Accordion News* magazine, corrected the error: "In a recent radio broadcast, Walter Winchell conveyed the information that Pietro Deiro had been married to Mae West for four years. As one Walter to another, I must set him right. Pietro was never married to the 'come up and see me sometime' girl. Guido Deiro, his brother, was supposed to be the fortunate accordionist." West made no public statements indicating that she had been married to Deiro. She referred to him simply as "D" in her autobiography. West's biographers state that the two never married. If they were married, this would have constituted bigamy as West was legally married to Frank Wallace at the time. West and Deiro split in 1916. Deiro's son claimed that years later Mae West privately revealed to him that she had become pregnant by Guido, and had an abortion without his knowledge, resulting in complications which left her sick for nearly a year and ultimately unable to bear children. According to Deiro's biographer, West filed for divorce on the grounds of adultery on July 14, 1920. The divorce was granted by the Supreme Court of the State of New York on November 9 of that year. West later said, "Marriage is a great institution. I'm not ready for an institution yet." Mae West remained close to her family throughout her life and was devastated by her mother's death in 1930. In that year, she moved to Hollywood and into the penthouse at the historic Ravenswood apartment building (where she would live until her death in 1980). After she began her movie career, her sister, brother and father followed her there. West provided them with nearby homes and also jobs and sometimes financial support.

Another person whom West spent her life with was attorney James Timony. She met Timony, who was fifteen years her senior, in 1916 when she was a vaudeville actress. They became romantically involved and he also began to act as her manager. By the mid-1930s when West was an established movie actress, they were no longer a couple. However, they remained extremely close, living in the same building, working together, and providing support for each other, until Timony's death in 1954. A year later, when she was 61, Mae West became romantically involved with one of the musclemen in her Las Vegas stage show, wrestler, former Mr. California, and former merchant marine Chester Rybonski (1923–1999). He was thirty years younger than West, and later changed his name to Paul Novak. He soon moved in with her and their romance continued until West died at the age of 87 in 1980. Novak once commented, "I believe I was put on this Earth to take care of Mae West." West also had many other boyfriends throughout her life. One was boxing champion William Jones, nicknamed Gorilla Jones. When the management at her apartment building discriminated against the African-American boxer and barred his entry, West solved the problem by buying the building.

Discography

Albums

- 1956: *The Fabulous Mae West*; Decca D/DL-79016 (several reissues up to 2006)

- 1960: *W.C. Fields His Only Recording Plus 8 Songs by Mae West*; Proscenium PR 22
- 1966: *Way Out West*; Tower T/ST-5028
- 1966: *Wild Christmas*; Dragonet LPDG-48
- 1970: *The Original Voice Tracks from Her Greatest Movies*; Decca D/DL-791/76
- 1970: *Mae West & W.C. Fields Side by Side*; Harmony HS 11374/HS 11405
- 1972: *Great Balls of Fire*; MGM SE 4869
- 1974: *Original Radio Broadcasts*; Mark 56 Records 643
- 1987/1995: *Sixteen Sultry Songs Sung by Mae West Queen of Sex*; Rosetta RR 1315
- 1996: *I'm No Angel*; Jasmine CD 04980 102
- 2006: *The Fabulous*: Rev-Ola CR Rev 181

At least 21 singles (78 rpm and 45 rpm) also were released from 1933 to 1973.

32

Florence Nightingale

Born in 1820 to a wealthy family, Florence was educated at home by her father. She aspired to serve others, in particular she wanted to become a nurse. Her parents were opposed - at that time, nursing was not seen as an attractive or 'respectable' profession. Despite her parents disapproval, Florence went ahead and trained to be a nurse. Florence later wrote that she felt suffocated by the vanities and social expectations of her upbringing. On one occasion, sitting in her parent's garden, she felt a call from God to serve others. She resolved to try and follow God's will in being of service to others. Florence had the opportunity to marry, but she refused a couple of suitors. She felt marriage would enslave her in domestic responsibilities. In 1853, the Crimea war broke out. This was a bloody conflict leading to many casualties on both sides. Reports of the British casualties were reported in the press; in particular it was noted that the wounded lacked even the most basic of first aid treatment. Many soldiers were dying unnecessarily. This was a shock to the British public, as it was one of the first wars to be reported vividly in the press back home. Later in 1855, Florence Nightingale was asked (with the help of her old friend Sydney Herbert) to travel to the Crimea and organise a group of nurses. Many of the initial applicants were unsuitable and Florence was strict in selecting and training the other nurses.

Florence was very gald to be able to take up the post and put into use her training as a nurse. They were based at the staff hospital at Scutari. She was overwhelmed by the primitive and chaotic conditions. There were insufficient beds for the men and conditions were terrible; the place smelt, was dirty, and even had rats running around spreading disease. Speaking of Scutari hospital, Florence Nightingale, said

> "The British high command had succeeded in creating the nearest thing to hell on earth."

In the beginning, the nurses were not even allowed to treat the dying men, they were only instructed to clean the hospital. But, eventually the number of casualties became so overwhelming the doctors asked Florence and her team of nurses to help. Florence's attitude included strict discipline for her other nurses, who always wore a highly visible uniform. The efforts of Florence and her team of nurses were greatly appreciated by the wounded soldiers and gradually positive news reports filtered back home. By the time she returned home she had become a national heroine and was decorated with numerous awards including one from Queen Victoria. After the war, she didn't really appreciate the fame, but continued to work for the improvement of hospital conditions, writing to influential people encouraging them to improve hygiene standards in hospitals. She also founded a training school for nurses at St Thomas's hospital, London. It was after her return from the Crimea that some of her most influential work occurred. She was a pioneer in using statistical methods to quantify the effect of different practises. Ironically, she found that some of her own methods of treating soldiers decreased recovery rates. But, this scientific approach to dealing with hospital treatment helped to improve standards and the quality of care. Florence Nightingale died at the age of 90 in 1910. Another nurse who gained a strong reputation at the time of the Crimean War was Jamaican nurse, Mary Seacole. Florence Nightingale didn't accept her offer of services when she came to the Crimea. But, Mary worked on her own initiative from a base in Balaclava near the front line. Her reputation amongst British officers was as strong as the reputation of Florence Nightingale. But, it was Florence Nightingale who remained in the public consciousness in the twentieth century.

33

Susan B. Anthony

Susan Brownell Anthony (February 15, 1820 – March 13, 1906) was a prominent American civil rights leader who played a pivotal role in the 19th century women's rights movement to introduce women's suffrage into the United States. She was co-founder of the first Women's Temperance Movement with Elizabeth Cady Stanton as President. She also co-founded the women's rights journal, *The Revolution*. She traveled the United States and Europe, and averaged 75 to 100 speeches per year. She was one of the important advocates in leading the way for women's rights to be acknowledged and instituted in the American government.

Early Life

Susan B. Anthony was born and raised in West Grove, Adams, Massachusetts. She was the second oldest of seven children—Guelma Penn (1818–1873), Hannah Lapham (1821–1877), Daniel Read (1824–1904), Mary Stafford (1827–1907), Eliza Tefft (1832–1834), and Jacob Merritt (1834–1900)—born to Daniel Anthony (1794–1862) and Lucy Read (1793–1880). One brother, publisher Daniel Read Anthony, would become active in the anti-slavery movement in Kansas, while a sister, Mary Stafford Anthony, became a teacher and a woman's rights activist. Anthony remained close to her sisters throughout her life. Her earliest American ancestors were the immigrants John Anthony (1607 - 1675), who was from Hempstead, Essex and his wife Susanna Potter (c. 1623 - 1674), who was from London, Middlesex. Anthony's father Daniel was a cotton manufacturer and abolitionist, a stern but open-minded man who was born into the Quaker religion. He did not allow toys or amusements into the household, claiming that they would distract the soul from the "inner light." Her mother, Lucy, was a student in Daniel's school; the two fell in love and agreed to marry in 1817, but Lucy was less sure about marrying into the Society of Friends (Quakers). Lucy attended the Rochester women's rights convention held in August 1848, two weeks after the historic Seneca Falls Convention, and signed the Rochester convention's Declaration of Sentiments. Lucy and Daniel Anthony enforced self-discipline, principled convictions, and belief in one's own self-worth. Susan was a precocious child, having learned to read and write at age three. In 1826, when she was six years old, the Anthony family moved from Massachusetts to Battenville, New York. Susan was sent to attend a local district school, where a teacher refused to teach her long division because of her gender. Upon learning of the weak education she was receiving, her father promptly had her placed in a group home school, where he taught Susan himself. Mary Perkins, another teacher there, conveyed a progressive image of womanhood to Anthony, further fostering her growing belief in women's equality. In 1837, Anthony was sent to Deborah Moulson's Female Seminary, a Quaker boarding school in Philadelphia. She was not happy at Moulson's, but she did not have to stay there long. She was forced to end her formal studies because her family, like many others, was financially ruined during the Panic of 1837. Their losses were so great that they attempted to sell everything in an auction, even their most personal belongings, which were saved at the last minute when Susan's uncle, Joshua Read, stepped up and bid for them in order to restore them to the family. In 1839, the family moved to Hardscrabble, New York, in the wake of the panic and economic depression that followed. That same year, Anthony left home to teach and pay off her father's debts.

She taught first at Eunice Kenyon's Friends' Seminary, and then at the Canajoharie Academy in 1846, where she rose to become headmistress of the Female Department. Anthony's first occupation inspired her to fight for wages equivalent to those of male teachers, since men earned roughly four times more than women for the same duties. In 1849, at age 29, Anthony quit teaching and moved to the family farm in

Rochester, New York. She began to take part in conventions and gatherings related to the temperance movement. In Rochester, she attended the local Unitarian Church and began to distance herself from the Quakers, in part because she had frequently witnessed instances of hypocritical behavior such as the use of alcohol amongst Quaker preachers. As she got older, Anthony continued to move further away from organized religion in general, and she was later chastised by various Christian religious groups for displaying irreligious tendencies. In her youth, Anthony was very self-conscious of her appearance and speaking abilities. She long resisted public speaking for fear she would not be sufficiently eloquent. Despite these insecurities, she became a renowned public presence, eventually helping to lead the women's movement.

Early Social Activism

Universal manhood suffrage, by establishing an aristocracy of sex, imposes upon the women of this nation a more absolute and cruel despotism than monarchy; in that, woman finds a political master in her father, husband, brother, son. The aristocracies of the old world are based upon birth, wealth, refinement, education, nobility, brave deeds of chivalry; in this nation, on sex alone; exalting brute force above moral power, vice above virtue, ignorance above education, and the son above the mother who bore him.

National Woman Suffrage Association

In the era before the American Civil War, Anthony took a prominent role in the New York anti-slavery and temperance movements. In 1836, at age 16, Susan collected two boxes of petitions opposing slavery, in response to the gag rule prohibiting such petitions in the House of Representatives. In 1849, at age 29, she became secretary for the *Daughters of Temperance*, which gave her a forum to speak out against alcohol abuse, and served as the beginning of Anthony's movement towards the public limelight. In late 1850, Anthony read a detailed account in the New York Tribune of the first National Women's Rights Convention in Worcester, Massachusetts. In the article, Horace Greeley wrote an especially admiring description of the final speech, one given by Lucy Stone. Stone's words catalyzed Anthony to devote her life to women's rights. In the summer of 1852, Anthony met both Greeley and Stone in Seneca Falls. In 1851, on a street in Seneca Falls, Anthony was introduced to Elizabeth Cady Stanton by a mutual acquaintance, as well as fellow feminist Amelia Bloomer. Anthony joined with Stanton in organizing the first women's state temperance society in America after being refused admission to a previous convention on account of her sex, in 1851. Stanton remained a close friend and colleague of Anthony's for the remainder of their lives, but Stanton longed for a broader, more radical women's rights platform. Together, the two women traversed the United States giving speeches and attempting to persuade the government that society should treat men and women equally.

Anthony was invited to speak at the third annual National Women's Rights Convention held in Syracuse, New York in September 1852. She and Matilda Joslyn Gage both made their first public speeches for women's rights at the convention. Anthony began to gain notice as a powerful public advocate of women's rights and as a new and stirring voice for change. Anthony participated in every subsequent annual National Women's Rights Convention, and served as convention president in 1858. In 1856, Anthony further attempted to unify the African-American and women's rights movements when, recruited by abolitionist Abby Kelley Foster, she became an agent for William Lloyd Garrison's American Anti-Slavery Society of New York. Speaking at the Ninth National Women's Rights Convention on May 12, 1859, Anthony asked *"Where, under our Declaration of Independence, does the Saxon man get his power to deprive all women and Negroes of their inalienable rights?"*

The Revolution

On January 8, 1868, Anthony first published the women's rights weekly journal *The Revolution*. Printed in New York City, its motto was: *"The true republic—men, their rights and nothing more; women, their rights and nothing less."* Anthony worked as the publisher and business manager, while Elizabeth Cady Stanton acted as editor. The main thrust of *The Revolution* was to promote women's and African-Americans' right to suffrage, but it also discussed issues of equal pay for equal work, more liberal divorce laws and the church's position on women's issues. The journal was backed by independently wealthy George Francis Train, who provided $600 in starting funds. His financial support ceased by May 1869, and the paper began to operate in debt. Anthony insisted on expensive, high-quality printing equipment, and she paid women workers the high wages she thought they deserved. She banned any advertisements for alcohol- and morphine-laden patent medicines; all such medicines were abhorrent to her. However, revenue from non-patent-medicine advertisements was too low to cover costs. In addition, Anthony got President Johnson to subscribe to the weekly journal before the first publication. In June 1870, Laura Curtis Bullard, a Brooklyn-based writer whose parents became wealthy from selling a popular morphine-containing patent medicine called "Mrs. Winslow's Soothing Syrup", bought the rights to *The Revolution* for one dollar, with Anthony assuming its $10,000 debt, an amount equal to $173,000 in current value. Anthony used her lecture fees to repay the debt, completing the task in six years. Under Bullard, the journal adopted a literary orientation and accepted patent medicine ads, but it folded in February 1872.

American Equal Rights Association

In 1869, long-time friends Frederick Douglass and Susan B. Anthony found themselves, for the first time, on opposing sides of a debate. The American Equal Rights Association (AERA), which had originally fought for both blacks' and women's right to suffrage, voted to support the 15th Amendment to the Constitution, granting suffrage to black men, but not women. Anthony questioned why women should support this amendment when black men were not continuing to show support for women's voting rights. Partially as a result of the decision by the AERA, Anthony soon thereafter devoted herself almost exclusively to the agitation for women's rights.

United States v. Susan B. Anthony

On November 18, 1872, Anthony was arrested by a U.S. Deputy Marshal for voting illegally in the 1872 Presidential Election two weeks earlier. She had written to Stanton on the night of the election that she had "positively voted the Republican ticket—straight...". She was tried and convicted seven months later, despite the stirring and eloquent presentation of her arguments that the recently adopted Fourteenth Amendment, which guaranteed to "All persons born or naturalized in the United States, and subject to the jurisdiction thereof, are citizens of the United States and of the State wherein they reside. No State shall make or enforce any law which shall abridge the privileges or immunities of citizens of the United States; nor shall any State deprive any person of life, liberty, or property, without due process of law; nor deny to any person within its jurisdiction the equal protection of the laws." the privileges of citizenship, and which contained no gender qualification, gave women the constitutional right to vote in federal elections. Her trial took place at the Ontario County courthouse in Canandaigua, New York, before Supreme Court Associate Justice Ward Hunt. Justice Hunt refused to allow Anthony to testify on her own behalf, allowed statements given by her at the time of her arrest to be allowed as "testimony," explicitly ordered the jury to return a guilty verdict, refused to poll the jury afterwards, and read an opinion he had written before the trial even started. The sentence was a $100 fine, but not imprisonment; true to her word in court ("I shall never pay a dollar of your unjust penalty"), she never paid the fine for the rest of her life, and an embarrassed U.S. Government took no collection action against her. The trial gave Anthony the opportunity to spread her arguments to a wider audience than ever before.

Anthony toured Europe in 1883 and visited many charitable organizations. She wrote of a poor mother she saw in Killarney that had "six ragged, dirty children" to say that "the evidences were that 'God' was about to add a No. 7 to her flock. What a dreadful creature their God must be to keep sending hungry mouths while he withholds the bread to fill them!" In 1893, she joined with Helen Barrett Montgomery in forming a chapter of the Woman's Educational and Industrial Union (WEIU) in Rochester. In 1898, she also worked with Montgomery to raise funds to open opportunities for women students to study at the University of Rochester.

National Suffrage Organizations

In 1869, Anthony and Elizabeth Cady Stanton founded

the National Woman Suffrage Association (NWSA), an organization dedicated to gaining women's suffrage. Anthony insisted that Stanton become president as long as possible; Anthony served as vice-president-at-large until 1892 when she became president. In the early years of the NWSA, Anthony made many attempts to unite women in the labor movement with the suffragist cause, but with little success. She and Stanton were delegates at the 1868 convention of the National Labor Union. However, Anthony inadvertently alienated the labor movement not only because suffrage was seen as a concern for middle-class rather than working-class women, but because she openly encouraged women to achieve economic independence by entering the printing trades, where male workers were on strike at the time. Anthony was later expelled from the National Labor Union over this controversy. In February 1890, Anthony orchestrated the merger of the NWSA with Lucy Stone's more moderate American Woman Suffrage Association (AWSA), creating the National American Woman Suffrage Association (NAWSA). This merger was partially done because Anthony admired Anna Howard Shaw, who worked with the AWSA and was a great speaker. Prior to the controversial merge, Anthony had created a special NWSA executive committee to vote on whether they should merge with the AWSA, despite the fact that using a committee instead of an all-member vote went against the NWSA constitution. Motions to make it possible for members to vote by mail were strenuously opposed by Anthony and her adherents, and the committee was stacked with members who favored the merger. (Two members who voted against the merger were asked to resign).

Anthony's pursuit of alliances with moderate suffragists created long-lasting tension between herself and more radical suffragists like Stanton. Stanton openly criticized Anthony's stance, writing that Anthony and AWSA leader Lucy Stone "see suffrage only. They do not see woman's religious and social bondage." Anthony responded to Stanton: "We number over ten thousand women and each one has opinions ... and we can only hold them together to work for the ballot by letting alone their whims and prejudices on other subjects!" The creation of the NAWSA effectively marginalized the more radical elements within the women's movement, including Stanton. Anthony pushed for Stanton to be voted in as the first NAWSA president, and stood by her as Stanton was belittled by the large factions of less-radical members within the new organization. In collaboration with Stanton, Matilda Joslyn Gage, and Ida Husted Harper, Anthony published *The History of Woman Suffrage* (4 vols., New York, 1884–1887). Anthony also befriended Josephine Brawley Hughes, an advocate of women's rights and Prohibition in Arizona, and Carrie Chapman Catt, whom Anthony endorsed for the presidency of the NAWSA when Anthony formally retired in 1900.

Later Personal Life, Death

Before retiring, Anthony was asked if all women in the United states would ever be given the vote. She replied by stating, "it will come, but I shall not see it...It is inevitable. We can no more deny forever the right of self-government to one-half our people than we could keep the Negro forever in bondage. It will not be wrought by the same disrupting forces that freed the slave, but come it will, and I believe within a generation." "Failure is impossible" were the words she left with her "girls" to encourage them on in the long discouraging struggle ahead. Fourteen years after Anthony's death, following assiduous campaigning, women were given the right vote on August 26, 1920, by the nineteenth amendment to the constitution. After retiring in 1900, Anthony remained in Rochester, where she died of heart disease and pneumonia in her house at 17 Madison Street on March 13, 1906. She was buried at Mount Hope Cemetery. Following her death, the New York State Senate passed a resolution remembering her "unceasing labor, undaunted courage and unselfish devotion to many philanthropic purposes and to the cause of equal political rights for women."

Legacy

Susan B. Anthony, who died 14 years before passage of the 19th Amendment giving women the right to vote, was honored as the first real (non-allegorical) American woman on circulating U.S. coinage with her appearance on the Susan B. Anthony dollar. The coin, approximately the size of a U.S.

quarter, was minted for only four years, 1979, 1980, 1981, and 1999. Anthony dollars were minted for circulation at the Philadelphia and Denver mints for all four years, and at the San Francisco mint for the first three production years. She was featured on a 3¢ U.S. commemorative stamp in 1936 and a 50¢ Liberty Issue regular issue stamp on August 25, 1955. The Susan B. Anthony List, a pro-life organization, is named in her honor. Anthony's position on abortion (or lack thereof) has been the subject of a long-running dispute. Anthony's birthplace in Adams was purchased in August 2006 by Carol Crossed, founder of the New York chapter of Democrats for Life of America and affiliated with Feminists for Life (FFL). The museum is operated independently of those two groups, though two FFL leaders serve on the museum's advisory board. Anthony's childhood home in Battenville, New York, was placed on the New York State Historic Register in 2006, and the National Historic Register in 2007. The Susan B. Anthony House in Rochester was declared a National Historic Landmark in 1966 and was operated as a museum. The American composer Virgil Thomson and poet Gertrude Stein wrote an opera, *The Mother of Us All*, that abstractly explores Anthony's life and mission. Along with Elizabeth Cady Stanton and Lucretia Mott, she is commemorated in *The Woman Movement*, a sculpture by Adelaide Johnson, unveiled in 1921 at the United States Capitol.

34

Emily Dickinson

General Summary Emily Dickinson

Emily Dickinson, regarded as one of America's greatest poets, is also well known for her unusual life of self imposed social seclusion. Living a life of simplicity and seclusion, she yet wrote poetry of great power; questioning the nature of immortality and death, with at times an almost mantric quality. Her different lifestyle created an aura; often romanticised, and frequently a source of interest and speculation. But ultimately Emily Dickinson is remembered for her unique poetry. Within short, compact phrases she expressed far-reaching ideas; amidst paradox and uncertainty her poetry has an undeniable capacity to move and provoke.

Early Life Emily Dickinson

Emily Dickinson was born on 10th December, 1830, in the town of Amherst, Massachusetts. Amherst, 50 miles from Boston, had become well known as a centre for Education, based around Amherst College. Her family were pillars of the local community; their house known as "The Homestead" or "Mansion" was often used as a meeting place for distinguished visitors including, Ralph Waldo Emerson. (although it unlikely he met with Emily Dickinson) As a young child, Emily proved to be a bright and conscientious student. She showed a sharp intelligence, and was able to create many original writings of rhyming stories, delighting her fellow classmates. Emily's father was strict and keen to bring up his children in the proper way. Emily said of her father. "his heart was pure and terrible". His strictness can be shown through his censorship of reading materials; Walt Whitman for example was considered "too inappropriate" and novels had to be smuggled into the house. In response, Emily was highly deferential to her father and other male figures of authority. But in her own way she loved and respected her father, even if at times, he appeared to be aloof. At a young age, she said she wished to be the "best little girl". However despite her attempts to please and be well thought of, she was also at the same time independently minded, and quite willing to refuse the prevailing orthodoxy's on certain issues.

Religious Influence on the Poetry of Emily Dickinson

A crucial issue at the time was the issue of religion, which to Emily was the "all important question" The antecedents of the Dickinson's can be traced back to the early Puritan settlers, who left Lincolnshire in the late 17th Century. Her antecedents had left England, so they could practise religious freedom in America. In the nineteenth- century, religion was still the dominant issue of the day. The East coast, in particular, saw a revival of strict Calvinism; developing partly in response to the more inclusive Unitarianism. Amherst College itself was founded with the intention of training ministers to spread the Christian word. Calvinism. By incrimination, Emily Dickinson would probably have been more at ease with the looser and more inclusive ideology of Unitarianism. However, the "Great Revival" as it was known, pushed the Calvinist view to greatest prominence.

Emily Dickinson's Seclusion

Because of her discomfort and shyness in social situations, Emily gradually reduced her social contacts, going out less and less into society. By her late twenties, this has led to an almost complete seclusion; spending most of her time in the family house, rarely meeting others from outside a close family circle. Her sister explains this wasn't a sudden decision, but a gradual process that happened over a period of time. However, despite the physical seclusion, Emily still maintained written contact with a variety of thought provoking people. It is also clear from her poetry that her decision to live life as a recluse did not close her mind, but in many ways allowed the flow of new avenues of thought and inner experiences. Despite her family's strong political tradition, Emily appeared unconcerned with politics. At the start of the American civil war she commented little on the event, and choose not to help the war effort, through making bandages. To be fair, this attitude of distancing from the war was quite common in the north. For example, her brother Austin choose to pay $500 to avoid military service; however as the war years advanced and Amherst experienced its first casualties of war, inevitably its citizens were drawn further into the conflict. Emily and her family, were particularly affected when friends of the family were killed in battle. Death of close friends was a significant feature of Emily's life; many close to her were taken away. This inevitably heightened her interest, fascination and perhaps fear of death, which informed so much of her poetry. The Civil War years were also the most productive for Emily; in terms of quantity of poems, it appears Emily Dickinson was influenced imperceptibly by the atmosphere of War, even if it appeared somewhat distant to her.

As well as writing over 1,700 poems, Emily was a prolific letter writer; these letters giving her the opportunity for contact with others, that in other respects she denied herself. Her letters show her love of language and are often not too dissimilar to her style of poetry. She went to great length to express her personal sentiments of gratitude and love to others. Her letters to her sister in law Sue have often been interpreted as love letters, leading to speculation over her sexual bias. But it must be remembered this emotional style of writing and communicating was fairly common for the time. They should also be seen in regard to Emily's other letters, which freely express intense emotional sentiments. Many of her poems refer to an invisible lover, - an object of devotion. Biographers have inevitably speculated about who this is. There is strong evidence that towards the end of her life she had some kind of emotional relationship with Judge Otis Lord (many years her senior and highly respected within the community). However, the poetry of Emily Dickinson was often deliberately vague.

The object of her devotion may have been no person in particular, but some unknown aspect of the divine. Emily Dickinson died at the age of 55 from Bright's disease, which is caused by kidney degeneration. Her doctor suggested that the accumulation of stress throughout her life contributed to her premature death. After her death, her close sister Vinnie, had been instructed to burn her letters. In doing so she came across a box of 1,700 of Emily's poems. Thankfully Vinnie ignored any request to burn old manuscripts. After a couple of years, Vinnie handed them to a family friend, Mabel Todd. Although Mabel had never met Emily, she had often been to Evergreens, the Dickinson family home. She typed up 200 letters becoming increasingly enthusiastic about the beauty and power of the poems. With the help and encouragement of Terrence Higginson, Emily's long standing friend, the first edition of poems was published in 1893. Her poems soon received extraordinary praise from leading magazines and newspapers. The New York Times claimed Emily Dickinson would soon be known amongst the immortals of English speaking poets.

35

Marie Antoinette

Marie Antoinette (2 November 1755 – 16 October 1793) was an Archduchess of Austria and the Queen of France and of Navarre. She was the fifteenth and penultimate child of Empress Maria Theresa of Austria and Emperor Francis I. In April 1770, on the day of her marriage to Louis-Auguste, Dauphin of France, she subsequently became Dauphine of France. Marie Antoinette assumed the title of Queen of France and of Navarre when her husband, Louis XVI of France, ascended the throne upon the death of Louis XV in May 1774. After seven years of marriage, she gave birth to a daughter, Marie-Thérèse Charlotte, the first of four children. Initially charmed by her personality and beauty, the French people generally came to dislike her, accusing "the Austrian" of being profligate and promiscuous, and of harboring sympathies for France's enemies, particularly Austria, since Marie Antoinette was, after all, Austrian. At the height of the French Revolution, Louis XVI was deposed and the monarchy abolished on 10 August 1792; the royal family was subsequently imprisoned at the Temple Prison. Nine months after her husband's execution, Marie Antoinette was herself tried, convicted of treason, and executed by guillotine on 16 October 1793. Maria Antonia of Austria was born on 2 November 1755 at the Hofburg Palace in Vienna, Austria; on the next day, she was baptised Maria Antonia Josepha Johanna (also known as Maria Antonia Josephina Johanna).

French Queen

The life of Marie Antoinette was the stuff of dreams when she was married at age 14 to the crown prince of France, the dauphin. France was then the most powerful nation of continental Europe, and the royal palace at Versailles the most opulent. The young princess could hardly have hoped for a more prestigious marriage and her magnificent marriage ceremony in 1770 was unmatched in royal pageantry. At the border she was stripped and re-dressed with clothing fashionable at the French court. When she was presented to the French king Louis XV, he pronounced her delightful, and told others of her fine full figure, of which he much approved. She became dauphine surrounded by all the comforts of the French court. Her enchanted life reached its pinnacle when the old king died and her husband became King Louis XVI in 1774. Marie Antoinette, still a teenager became Queen of France.

Unhappy Marriage and Boredom

But this daughter of life's fortune was unhappy in her marriage. Louis was homely, awkward and hardly her heart's desire. His devotion to the hunt, clocks and his workshop and his early hours were in contrast to her pursuit of the arts, fashion, dance and French nightlife. The contrast of Charles and Diana comes to mind. While King Louis XV, her husband's brothers, Provence and Artois, and others at court noticed at once her grace and beauty, her own shy husband was slow to exercise the rights of the marriage bed. From afar, Louis XVI, like the others, much admired Marie's physical charms and her character, and Louis would become a thoroughly devoted husband, but in her early years in France he was little comfort to her. Pushed by her mother's letters, Marie still sought out Louis. Yet, to add to Antoinette's frustration, even when she could achieve intimacy with him, Louis was unable to achieve erection. So, Antoinette and Louis were unable to have sex and their marriage went unconsummated for seven years. It took the intervention of the Queen's oldest brother, emperor Joseph of Austria, in a heart to heart meeting with Louis in 1777, to convince him to have the needed operation. Meanwhile, the teenage queen suffered in silence as she was snidely taunted for her inability to produce an heir to the throne. Beyond her personal frustrations with her husband, Marie Antoinette was bored with her position and its duties. The days of the young princess and then queen were spent in endless court rituals and strict etiquette tracing to the days of Louis XIV. The young queen tired of being constantly on public display with the requirements of her position. She missed the more relaxed environment and freedom of Vienna. Her displeasure and sarcasm directed at the older aunts and members of the high nobility were noticed and commented upon.

Friends Circle

Marie Antoinette sought escape from her marital frustration and the boredom of court life. Time went by and she began to exercise power as queen, Marie Antoinette spent less time at court, and surrounded herself with a dissolute clique, led by Yolande de Polignac and Thérèse de Lamballe. She lavished expensive gifts and positions upon these friends and in doing so ignored the great houses of the French nobility. With her young friends, Marie Antoinette threw herself into a life of pleasure and careless extravagance. These included masked balls in Paris, gambling, theatricals and late night promenades in the park. Her circle included the King's frivolous young brother the Count of Artois, and handsome young courtiers the Duc de Ligne, Counts Dillon, Vaudreuil and Axel Fersen. The Queen's indiscretions with her circle of friends led to scandals such as the Diamond Necklace Affair and rumours concerning her relations with that circle including Axel Fersen.

Extravagant Life

The young queen, with her blonde beauty and style set fashion trends through France and Europe. Her painter Vigee Lebrun commented about the translucent colour of her complexion, her long blonde hair and her well-proportioned and full-bosomed figure. All commented how well she carried herself. Her page Tilly said she walked better than any woman and as you'd offer a woman a chair, you'd offer her a throne. The queen enjoyed her beauty style, but her fashion fame came at a price. The Queen spent lavishly on her dress and adornments. Each year she exceeded her clothing allowance which the King covered. The excessive fashions for high headdresses, plumes and voluminous dresses were subject to public comment, caricature and on occasion ridicule. The queen also spent lavishly on her friends as mentioned and on her entertainment including her retreat at Petit Trianon. This small palace adjoining Versailles was given to Marie by Louis XVI. There she arranged extensive interior decorations and building of a theatre for her theatricals and the Temple of Love in the park. Marie also had built a rustic Viennese retreat called the hameau. Here, she played at being at being a simple milkmaid. To add to the fun, Sevres porcelain bowls were cast using Marie Antoinette's own ample breasts as their mould (as was said to have had been done in the case of Helen of Troy). The hameau was stocked with perfumed sheep and goats, but the actual milking and chores were done by servants.

Anger at the Queen

By the late 1780s, envy and hatred of Marie Antoinette were widespread. Many at court had always opposed the Austrian alliance, and had resented her efforts to intercede on occasion for Austrian causes. The king's brother the Count of Provence and his cousin the Count of Orleans both thought they were more capable than Louis XVI. They were jealous both of Louis's kingship and his marriage to the beautiful Marie Antoinette. Many others among the nobility were envious of the Queen and insulted by her dismissal of court etiquette, preference for her small court circle and the patronage she wielded on their behalf. Thus, disaffected members of the nobility became fertile sources for dirt on the queen. They fabricated and circulated scurrilous stories about the Queen and her private life. Stories accused of all sort of sexual acts with men and women of the court, of sending funds to Austria, and challenged the paternity of the royal children.

Diamond Necklace Affair

By the mid 1780s tales of the queen's extravagance, dissipation and sexual vice abounded. It was at this point that the Diamond Necklace Affair became the sensation, grabbing the attention of the entire nation. The affair fused three disparate situation, united by widely held beliefs in the loose morals of Marie Antoinette. For years an impoverished scion of past Valois nobility, Madame Lamotte schemed to gain a position at court. At the same time, socially prominent Prince de Rohan, the Cardinal of France was unhappy over his years of exclusion from Marie Antoinette's inner circle, and the jeweller Boehmer was unable to convince Marie Antoinette to buy a fabulously expensive diamond necklace originally made for Louis XV's lover Madame du Barry. Lamotte was a full figured attractive woman who caught the attention of both men, and was able to convince them she was a lesbian lover of Marie Antoinette. Lamotte convinced Rohan that the Queen indeed wanted the necklace and Rohan obtained it from Boehmer and gave it to Lamotte after meeting a prostitute dressed as Marie Antoinette at a late night rendezvous near the Temple of Love, where the Queen was said to hold lovers' trysts with others. When Boehmer approached the Queen for payment (just as she was preparing for to play a role in a banned Beaumarchais play Le Figaro), the charade unravelled. When they learned the basic facts of the affair, both king and queen were enraged that Rohan would think that the queen would use a go-between to obtain a necklace.

Necklace Trial and Impact

Royal pique proved disastrous. The cardinal, highest churchman in France, was arrested on the Day of Assumption in the middle of the entire court. Next the Queen demanded public vindication, so the king obtained a trial before the Parlement of Paris. The trial proved a sensation for months, with the dirty laundry of the monarchy paraded before all France. The cast included the highest nobles, charlatans, a prostitute who looked like the Queen, and above all the fabulous diamond necklace and the Queen herself despite never being called as witness. In the end, the nobility displayed their defiance before the entire nation in the Diamond Necklace Affair with their acquittal of Prince de Rohan on the charge of insulting the queen. The ruling of the Parlement of Nobles effectively said that at the least, given her reputation, the queen was worthy of such insult. Rohan could reasonably believe Marie Antoinette would use him as a go-between and in the end exchange her sexual favours for a diamond necklace. When the not guilty verdict was announced in the crowded Paris opera house an enormous roar went up and all eyes turned to the royal box. A shocked Marie Antoinette hastily departed for her coach, amid the crowd's hoots. The court did convict the less well connected Lamotte, and she was branded on her breasts and imprisoned. But her husband had escaped to England and she escaped prison. She exacted her revenge by concocting and circulating a tale that she was indeed the queen's lesbian lover, that the queen was insatiable in her desires and that the queen got the necklace and the affair was all for her amusement. As fabulous as her story was, it circulated in the thousands and was widely believed. So much so that had she not died in 1793, Lamotte might well have testified against Marie at her trial.

Madame Deficit and Financial Crisis

Ironically, as the Diamond Necklace Affair erupted and the Queen's popularity sank to its nadir, age and maturity tempered her lifestyle. Louis and Marie were able to have children and Antoinette bore four children. She spent less time with Paris night life and more with her children and family. Though still graceful and attractive, as she passed age thirty, Marie's increasingly stout figure moved her toward darker colours. Her milliner Madame Bertin used less ostentatious fashion, while still showing Marie's large bust to fine advantage. Even as she still flirted with men of court and spent much time with Axel Fersen, Louis was increasingly devoted to his handsome wife whom he adored. While Marie's personal life was settling down, the state of France was not. France also had bad harvests in the late 1780s and the poor suffered. The Queen was good hearted and kindly and tried to aid the poor of her country. She attended benefits for charity (including the night the Necklace verdict was announced), and used the hameau to aid a number of impoverished families. However, her small acts were hardly noticed amid the suffering. What was remembered was that the queen played at being a milkmaid and shepherdess, at the manicured hameau of Trianon, while real peasants starved. Her perceived insensitivity led many to believe she said "Let them eat cake", when told of the widespread starvation. Furthermore, France reeled under huge debts inherited from Louis XV which Louis XVI had been unable to repay. France's debt was now a crisis, with the final straw being its France's costly aid from 1778 to 1783 to the American colonies in their War of Independence with Great Britain. To try to revive the Queen's popularity and rally support for the monarchy portraits were made and exhibited showing the Queen surrounded by her loving children. Yet the obvious royal propaganda backfired as detractors noticing the Queen's expansive costume, dubbed the pictured heroine, "Madame Deficit". It was at this time, amid such increased unpopularity and still reeling from the aftershocks of the Necklace Affair, when Louis XVI most needed support from the nobility. He tried to effect needed reforms through a series of ministers, relying in each instance on advice from his Queen, and then he called an assembly of notables to again try to effect reforms to deal with the financial crisis. Louis was not a forceful king, his wife's influence was resented and the position of the monarchy weakened.

Estates General – 1789

Tragedy struck Louis and Marie in 1789. Their oldest son and heir, the dauphin, was dying of a crippling, agonizing hereditary disease and would die in June. Besides her miscarriages, this was the second child dead; their second daughter had died in 1786. And now amid this grief, the couple faced the crisis that now threatened their rule, which would bring still further tragedies to this family. Unable to force the nobility to make needed financial reforms, the desperate king called the Estates General in May 1789. This was the first time in 175 years it was called. But it was unique because it gave representation to common men, as one of the three estates able to vote. Louis did this to try to gain the support of the common people (third estate) to force needed reforms. The Estate General did not begin auspiciously as the Queen's appearance was met first by silence and then call Vive Duc Orleans – her scorned suitor and hated foe. This rebelliousness was a sign of what was to follow. The common people were not content with the limited role of the third estate Louis envisioned. The genie was now out of the bottle. The third estate declared itself the national assembly and in the Tennis Court Oath said it would not adjourn until France had a constitution.

Fall of the Bastille

Louis lacked the will to quell this rebellion but was repeatedly lobbied to take action by Marie Antoinette. The queen strongly desired to preserve absolute monarchy and was firm in her opposition to reforms that would give greater power to the common people. However, with a taste of success, the common people did not want to see the third estate suppressed. In July, a mob of commoners seized the Invalides and obtained a supply of fire arms. The next effort was to obtain powder so they could defend the assembly as needed. For this effort the mob attacked a great symbol of absolute monarchy, the ancient and famous Bastille prison and fortress that loomed in the centre of Paris. Louis failed to take prompt action and the mob succeeded in taking the Bastille. The governor of the Bastille who resisted and threatened to blow up the gun powder was hacked to death by the mob his head sported on pikes for all to see. The crowd had arms and ammunition. Lawlessness had occurred and no royal action had been taken in response. Louis went to Paris to restore calm but no actions were taken against those who stormed the Bastille.

The Great Fear

The storming of the Bastille greatly disturbed a number of nobles who knew the poverty of the common people and feared vengeance if royal power was inadequate to check mob impulses. Leading members of the royal court, including close friends of Marie Antoinette fled the country. These included in July and August the Count of Artois and Madame Polignac and in October her close friend and portraitist Vigee Lebrun. The royal court at Versailles was just 20 miles from the raging cauldron of Paris. Marie Antoinette too feared the Paris mob and counselled Louis to repair to the country so he could quell rebellion from afar, but Louis would not leave Versailles.

The Queen's Fate

After her husband's death, in July 1793, Marie Antoinette's son was forcibly taken from her. The poor woman begged that her son be allowed to stay but she was powerless to change the will of the ministers. The boy was put under the care of Simon, a cobbler and one of the Commissaires of the Commune, and died of neglect within two years. In September 1793, Marie Antoinette was separated from her daughter and sister in law. Now called "Widow Capet", Marie was transferred to months of solitary confinement in the dank Conciergerie prison, where she was under twenty-four hour guard by revolutionaries who from behind their screen watched her every move. The Conciergerie prison was the antechamber to death. In this dank prison, she lost much weight and her eyesight began to fail, but she did not have long to live. On October 14, the poor pallid woman was awoken at night and faced the Revolutionary Tribunal. The trial was a horror, with the Queen attacked more as a person than as a queen. Her own son was forced to testify that she abused him. The queen bravely replied to all charges and to this she said, "If I make no reply, it is because I cannot, I appeal to all mothers in this audience."

Titles from Birth to Death

- 2 November 1755 – 19 April 1770: *Her Royal Highness* Archduchess Maria Antonia of Austria
- 19 April 1770 – 10 May 1774: *Her Royal Highness* The Dauphine of France
- 10 May 1774 – 1 October 1791: *Her Majesty* The Queen of France
- 1 October 1791 – 21 September 1792: *Her Majesty* The Queen of the French
- 21 January 1793 – 16 October 1793: Widow Capet

36

Emily Pankhurst

> "Human life for us is sacred, but we say if any life is to be sacrificed it shall be ours; we won't do it ourselves, but we will put the enemy in the position where they will have to choose between giving us freedom or giving us death."
>
> —*Emmeline Pankhurst From Freedom or Death (1913)*

Emily Pankhurst was born in Manchester in 1858. Her family had a tradition of radical politics and she stepped into that mould becoming a passionate campaigner for women's right to vote. She married Richard Pankhurst who supported the women's suffrage movement and his death in 1898, was a great shock to Emily. After his death, she threw herself into the women's suffrage movement forming the Women's Franchise league in 1898. In 1903 she formed the more militant Women's Social and Political Union (WSPU). It was through the WSPU that the political action gained the group the term women's suffragette movement. She led a passionate group of women who were willing to take part in drastic action such as tying to railings, smashing windows and launching demonstrations. The government and establishment were somewhat shocked at the tactics of the women and many were arrested. When they went on hunger strike they were force fed or released only to be rearrested - something known as cat and mouse. In 1912 Emily Pankhurst was convicted or breaking windows and sent to Holloway Prison. In prison she went on hunger strike in protest about the appaling conditions, prisoners were kept in.

She described her time in prison. "*like a human being in the process of being turned into a wild beast*" Before, the First World War, the women's suffrage movement gained increased exposure polarising public opinion. It was in 1913 that Emily Davison was killed when throwing herself under the King's horse. However, at the outbreak of war in 1914, Emily Pankhurst used her campaigning tactics to support the war effort - announcing a temporary truce in the women's suffrage campaign. She considered the menace of German aggression to be greater. The government and the suffragettes declared a truce and political prisoners were released. In the war effort, women were drafted into factories and took on many jobs previously the preserve of men such as bus drivers and postmen. The radical social change of the first world war helped to diminish the opposition to women getting the vote and in 1918, women over the age of 30 were given the vote. In 1929, the voting age for women was reduced to the same age as men. Emily died shortly after her life's goal was achieved.

37

Marie Curie

Early Life

Marya Sklodovska was the youngest of 5 children, born in 1867, Warsaw Poland. She was brought up in a poor but well educated family. Marya excelled in her studies and won many prizes. At an early age she became committed to the ideal of Polish independence from Russia which was currently ruling Poland with an iron fist, and in particular making life difficult for intellectuals. She yearned to be able to teach fellow Polish woman who were mostly condemned to zero education. Unusually for women at that time, Marya took an interest in Chemistry and Biology. Since opportunities in Poland for further study was limited, Marya went to Paris, where after working as a governess she was able to study at the Sorbonne, Paris. Struggling to learn in French, Marya threw herself into her studies, leading an ascetic life dedicated to studying. She went on to get a degree in Physics finish top in her school. She later got a degree in Maths, finishing second in her school year. It was in Paris, that she met Pierre Curie, who was then chief of the laboratory at the school of Physics and Chemistry. He was a renowned Chemist, who had conducted many experiments on crystals and electronics. Pierre was smitten with the young Marya and asked her to marry him. The unromantic Marya initially refused, but, after persistence from Pierre she relented. The two would later become inseparable, until Pierre's untimely death.

Marie Curie Work on Radioactivity

Marie pursued studies in radioactivity. In 1898, this led to the discovery of two new elements. One of which she named polonium after her home country. There then followed 4 years of extensive study into the properties of radium. Using dumped uranium tailings from a nearby mine, they were very slowly, and painstakingly, able to extract a decigram of radium. Radium was discovered to have remarkable impacts. Marie actually suffered burns from the rays. It was from this discovery of radium and its properties that the science of radiation was able to develop. Using the properties of radium to burn away diseased cells in the body. Initially radiotherapy was called 'currietherapy' The Curries agreed to give away their secret freely; they did not wish to patent such a valuable element. The element was soon in high demand and it began industrial scale production. For their discovery they were awarded the Davy Medal (britain) and the Nobel Prize for physics in 1903.

In 1905, Pierre was killed in a road accident, leaving Marie to look after the laboratory and her 2 children. In 1911 she was awarded a second nobel prize in Chemistry for the discovery of actinium and further studies on radium and polonium. The success of Marie Curie also brought considerable hostility, criticism and suspicion from a male dominated science world. She suffered from the malicious rumours and accusations that flew around. The onset of World War I in 1914, led to Marie Curie dedicating her time to the installation of X ray machines in hospitals. Marie understood that x ray machines would easily be able to located shrapnel, enabling better treatment for soldiers. By, the end of the first world war, over a million soldiers had been examined by her X ray units. At the end of the First world war she returned to the Institute of Radium in Paris, and also took great pride in serving the fledgling League of Nations. She also published a book - radioactivity which encompassed her great ideas on science. Marie Curie died in 1934 from Cancer. It was an unfortunate side effect of her own groundbreaking studies into radiation which were to help so many people. Marie Curie pushed back many frontiers in science; and at the same time set a new bar for female academic and scientific achievement.

38

Emily Murphy

Emily Murphy (born Emily Gowan Ferguson; 14 March 1868 – 17 October 1933) was a Canadian women's rights activist, jurist, and author. In 1916, she became the first woman magistrate in Canada, and in the British Empire. She is best known for her contributions to Canadian feminism, specifically to the question of whether women were "persons" under Canadian law. In 1927, Murphy and four other women: Henrietta Muir Edwards, Nellie McClung, Louise McKinney and Irene Parlby, who together came to be known as "The Famous Five" (also called "The Valiant Five"), launched the "Persons Case," contending that women could be "qualified persons" eligible to sit in the Senate. The Supreme Court of Canada ruled that they were not. However, upon appeal to the Judicial Committee of the British Privy Council, the court of last resort for Canada at that time, the women won their case.

Early Life

Emily Murphy was born the third of six children in Cookstown, Ontario to wealthy landowner and businessman Isaac Ferguson and his wife – also named Emily. As a child, Murphy frequently joined her two older brothers Thomas and Gowan in their adventures; in fact, their father encouraged this behaviour and often had his sons and daughters share responsibilities equally. Considering her family involvement in the law and politics, it is no surprise that Murphy became one of the most influential suffragists in Canada. Murphy grew up under the influence of her maternal grandfather, Ogle R. Gowan who was a politician that founded a local branch of the Orange Order in 1830 and two uncles who were a Supreme Court justice and a Senator, respectively. Her brother also became a lawyer and another member of the Supreme Court. Her family were prominent members of society and she benefited from parents who supported their daughter receiving formal academic education. She is as well related to James Robert Gowan, who was a lawyer, judge, and senator. Murphy attended Bishop Strachan School, an exclusive Anglican private school for girls in Toronto and, through a friend, she met her future husband Arthur Murphy who was 11 years her senior. In 1887, they were married and had four daughters Madeleine, Evelyn, Doris and Kathleen. Doris died young of diphtheria. After Doris' death, the family decided to try a new setting and moved west to Swan River, Manitoba in 1903 and then to Edmonton, Alberta in 1907.

Dower Act

While Arthur was working as an Anglican priest, Murphy explored her new surroundings and became increasingly aware of the poverty that existed. At the age of 40, when her children became independent and began their separate lives, Murphy began to actively organize women's groups where the isolated housewives could meet and discuss ideas and plan group projects. In addition to these organizations, Murphy began to speak openly and frankly about the disadvantaged and the poor living conditions that surrounded their society. Her strong interest in the rights and protection of women and children intensified when she was made aware of an unjust experience of an Albertan woman whose husband sold the family farm; the husband then abandoned his wife and children who were left homeless and penniless. At that time, property laws did not leave the wife with any legal recourse. This case motivated Murphy to create a campaign that assured the property rights of married women. With the support of many rural women, Murphy began to pressure the Alberta government to allow women to retain the rights of their land. In 1916, Murphy successfully persuaded the Alberta legislature to pass the Dower Act that would allow a woman legal rights to one third of her husband's property. Murphy's reputation as a women's rights activist was established by this first political victory.

The Persons Case

Murphy's success in the fight for the Dower Act, along with her work through the Local Council of Women and her increasing awareness of women's rights, influenced her request for a female magistrate in the women's court. In 1916, Murphy, along with a group of women, attempted to observe a trial for women who were labelled prostitutes and were arrested for "questionable" circumstances. The women were asked to leave the courtroom on the claims that the statement was not "fit for mixed company". This outcome was unacceptable to Murphy and she protested to the provincial Attorney General. "If the evidence is not fit to be heard in mixed company," she argued, "then the government must set up a special court presided over by women, to try other women." Murphy's request was approved and she became the first woman police magistrate for the British Empire. Her appointment as judge, however, became the cause for her greatest adversity concerning women within the law. In her first case in Alberta on 1 July 1916, she found the prisoner guilty. The prisoner's attorney called into question her right to pass sentence, since she was not legally a person. The Provincial Supreme Court denied the appeal. In 1917, she headed the battle to have women declared as "persons" in Canada, and, consequently, qualified to serve in the Senate. Lawyer, Eardley Jackson, challenged her position as judge because women were not considered "persons" under the British North America Act 1867. This understanding was based on a British Common Law ruling of 1876, which stated, "women were eligible for pains and penalties, but not rights and privileges."

Murphy began to work on a plan to ask for clarification of how women were regarded in the BNA act and how they were to become Senators. In order for her question to be considered, she needed at least five citizens to submit the question as a group. She enlisted the help of four other Albertan women and on 27 August 1927 she and human rights activist Nellie McClung, ex MLA Louise McKinney, women's rights campaigners Henrietta Edwards and Irene Parlby signed the petition to the Supreme Court of Canada. This women asked, "Does the word 'person' in Section 24 of the British North

America Act include female persons?" The campaign became known as The Persons Case and reached the Supreme Court of Canada on March 1928. The court denied the women from challenging the interpretation of the word "persons", which led the five women to bring the case to the Judicial Committee of the Privy Council in Britain. On 18 October 1929, in a decision called Edwards v. Canada (Attorney General), the Privy Council unanimously declared that women were also considered as "persons" under the BNA Act and were eligible to serve in the Senate. The women were known as the Famous Five and were considered leaders in education for social reform and women's rights. They challenged convention and established an important precedent in Canadian history. In Canada's Senate Chamber, the five women are honoured with a plaque that reads, "To further the cause of womankind these five outstanding pioneer women caused steps to be taken resulting in the recognition by the Privy Council of women as persons eligible for appointment to the Senate of Canada." Murphy, along with the rest of the Famous Five are featured on the back of the new Canadian 50 dollar bill. In October 2009, the Senate voted to name Murphy and the rest of the Five Canada's first "honorary senators."

Drugs and Race

Although Murphy's views on race changed over the course of her life, the perspective contained in her book, the *Black Candle*, is considered the most consequential because it played a role in creating a widespread "war on drugs mentality" leading to legislation that "defined addiction as a law enforcement problem." A series of articles in *Maclean's* magazine under her pen name, "Janey Canuck," forms the basis of the *Black Candle*. Using extensive anecdotes and "expert" opinion, the *Black Candle* depicts an alarming picture of drug abuse in Canada, detailing Murphy's understanding of the use and effects of opium, cocaine, and pharmaceuticals, as well as a "new menace," "marihuana." Murphy's concern with drugs began when she started coming into "disproportionate contact with Chinese people" in her courtroom because they were over represented in the criminal justice system. In addition to professional expertise and her own observations, Murphy was also given a tour of opium dens in Vancouver's Chinatown by local police detectives. Vancouver at the time was in the midst of a moral panic over drugs that was part of the anti-Oriental campaign that precipitated the Chinese Immigration Act of 1923. Canadian drug historian Catherine Carstairs has argued that Murphy's importance regarding drug policy has been "overstated" because she did not have an impact on the drug panic in Vancouver, but that nevertheless "her articles did mark a turning point and her book ... brought the Vancouver drug panic to a larger Canadian audience."

Race permeates the *Black Candle*, and is intricately entwined with the drug trade and addiction in Murphy's analysis. Yet she is ambiguous in her treatment of non-whites. In one passage, for example, she chastises whites who use the Chinese as "scapegoats," while elsewhere, she refers to the Chinese man as a "visitor" in this country, and that "it might be wise to put him out" if it turns out that this visitor carries "poisoned lollipops in his pocket and feeds them to our children." Drug addiction, however, not the Chinese immigrant, is "a scourge so dreadful in its effects that it threatens the very foundations of civilization," and which laws therefore need to target for eradication. Drugs victimize everyone, and members of all races perpetrate the drug trade, according to Murphy. At the same time, she does not depart from the dominant view of middle class whites at the time that "races" were discrete, biologically determined categories, naturally ranked in a hierarchy. In this scheme, the white race was facing degradation through miscegenation, while the more prolific "black and yellow races may yet obtain the ascendancy" and thus threatened to "wrest the leadership of the world from the British." Murphy's ambiguity regarding non-whites is reflected in scholarly debates, but what is not controversial is that the *Black Candle* was written "for the express purpose of arousing public demands for stricter drug legislation" and that in this she was to some degree successful. This motivation may have influenced her racial analysis by playing to the popular prejudices of her white audiences.

The Eugenics Movement

During the early twentieth century, scientific knowledge

emerged in the forefront of social importance. Advances in science and technology were thought to hold answers to current and future social problems. Murphy was among those who thought that the problems that were plaguing their society, such as alcoholism, drug abuse and crime were caused because of mental deficiencies. In a 1932 article titled " Overpopulation and Birth Control", she states: "... over-population [is a] basic problem of all...none of our troubles can even be allayed until this is remedied." As the politics behind the Second World War continued to develop, Murphy, who was a pacifist, theorized that the only reason for war was that nations needed to fight for land to accommodate their growing populations. Her argument was that: if there was population control, people would not need as much land. Without the constant need for more land, war would cease to exist.

Legacy

Her legacy is disputed, with her important contributions to feminism being weighed against her nativist views. In addition to being against immigration, she was a strong supporter of Alberta's legislation for the *Sexual Sterilization of the Insane* at a time when compulsory sterilization was practiced in some North American jurisdictions. However, it has been argued that that those in the vanguard make mistakes: Murphy's views were a product of her times. Recent memorializing of the Famous Five, such as the illustration on the back of the fifty dollar bill, has been used as the occasion for re-evaluating Murphy's legacy. Marijuana decriminalization activists especially have targeted Murphy for criticism as part of the movement to discredit marijuana prohibition. They charge that today's drug laws are built on the racist foundations laid by Murphy and that the drug war has harmed more women than the Persons Case has benefited. Conversely, Murphy's defenders have been quick to point out that she was writing at a time when white racism was typical, not exceptional, and that Murphy's views were more progressive than many of her peers. Moreover, her views on race or drugs in no way negate Murphy's positive accomplishments in advancing the legal status of women.

39

Rosa Luxemburg

Rosa Luxemburg (5 Mar 1870-15 Jan 1919), she was educated in Warshw and in 1886 she joined the Proletariat Party. She organised strikes and in 1889 she fled to Switzerland to avoid an arrest. In Zürich she studied philosophy, history, mathematics and economics.In 1898 she married Gustav Lübeck. She moved to Berlin, where she became an active member of the SPD. She published analyses in newspapers and she was strongly opposed to German militarism. She was imprisoned several times between 1904 and 1906. In 1907 she met Lenin in London. After the SPD refused a general strike in 1910 she broke with Karl Kautsky.She taught Marxism at the SPD training center in Berlin. Among her students was the future president Friedrich Ebert. In 1912 she represented the SPD at European events during the time that Jean Jaurès represented the French socialists.When the SPD promised the government to refrain from strikes during the war she organised demonstrations against the war in Frankfurt for which she was imprisoned for a year. After her release she collaborated with Karl Liebknecht, Clara Zetkin and Franz Mehring on pamphlets against the war that they signed "Spartacus". Rosa called herself "Junius".In June 1916 she was imprisoned for two and a half years. She was held prisoner in Poznan and Wroclaw, but friends smuggled her writings out of the prison and they were sunsequently published. She was release on 6 Nov 1918 and immediately founded the newspaper "Die Rote Fahne" ("The Red Flag") with Karl Liebknecht. In December 1918 se was one of the founders of the Communist Party KPD. Luxemburg and Liebknecht were arrested in Berlin on 15 Jan 1919 by a Freikorps under the command of Waldemar Pabst. They were questioned and subsequently killed. Rosa's body was thrown into the Landwehr Canal and her corpse wasn't found until 1 Jun 1919. In later years Pabst claimed that Ebert had agreed with the murders.

40

Zsa Zsa Gabor

Zsa Zsa Gabor (born February 6, 1917) is a Hungarian-born American actress on stage, film and television. She acted on stage in Vienna, Austria, in 1932, and was crowned Miss Hungary in 1936. She emigrated to the United States in 1941 and became a sought-after actress with "European flair and style", with a personality that "exuded charm and grace". Her first movie role was as supporting actress in *Lovely to Look At*. She later acted in *We're Not Married!* and played one of her few leading roles in *Moulin Rouge* (1952), directed by John Huston, who described her as a "creditable" actress. Besides her film and television appearances, she is best-known for having nine husbands, including hotel magnate Conrad Hilton and actor George Sanders. She once stated, "Men have always liked me and I have always liked men. But I like a mannish man, a man who knows how to talk to and treat a woman—not just a man with muscles."

Early Life and Career

Born as Sári Gábor in Budapest (then part of the Austro-Hungarian Empire), the middle of the three daughters of Vilmos Gábor (1884–1962), a soldier, and Jolie Gábor (1894–1997). Her elder sister Magda was a socialite and her younger sister Eva was an actress and businesswoman. Gabor's mother, Jolie (née Tillemann Jánosné), was a cousin of Annette Tilleman Lantos, the wife of Hungarian-born U.S. congressman and Holocaust survivor, Tom Lantos. Jolie was of Jewish descent and barely escaped from Hungary after the Nazis occupied Budapest in 1944. She credits Magda's husband for helping her: "For Magda's Portuguese Ambassador I thank God. It was this man who saved my life." Gabor's maternal grandparents chose to remain in Budapest feeling they "had a good place to hide." However, the U.S. later bombed Nazi positions in Budapest near where her grandparents were in hiding, and they both died during one of the bombing raids. Following studies at Madame Subilia's, a Swiss boarding school, Zsa Zsa Gabor was discovered by the tenor Richard Tauber on a trip to Vienna in 1936 and was invited to sing the soubrette role in his new operetta *Der singende Traum* ("The Singing Dream") at the Theater an der Wien, her first stage appearance. Zsa Zsa is unique. She's a woman from the court of Louis XV who has somehow managed to live in the 20th century, undamaged by the PTA ... She says she wants to be all the Pompadours and Du Barrys of history rolled into one, but she also says, 'I always goof. I pay all my own bills ... I want to choose the man. I do not permit men to choose me.' Television host Merv Griffin, in his autobiography, described the Gabors, "in their heyday," as "glamour personified": "All these years later, it's hard to describe the phenomenon of the three glamorous Gabor girls and their ubiquitous mother. They burst onto the society pages and into the gossip columns so suddenly, and with such force, it was as if they'd been dropped out of the sky." A biopic is to be made on her life by Italian director Gabriela Tagliavini who claimed that Gabor "is a perfect celebrity to be the focus of a movie". According to *Insider*, Gabor is "an original. Her free spirit, eccentricity and wicked wit made her one of the most memorable celebrities of our time." Gabor's husband will reportedly be involved in the film's production.

Personal Life

Gabor has been married nine times. She was divorced seven times, and one marriage was annulled. Her husbands, in chronological order, are:

- Burhan Asaf Belge (1937–1941) (divorced)
- Conrad Hilton (April 10, 1942–1947) (divorced)
- George Sanders (April 2, 1949 – April 2, 1954) (divorced)
- Herbert Hutner (November 5, 1962 – March 3, 1966) (divorced)
- Joshua S. Cosden, Jr. (March 9, 1966 – October 18, 1967) (divorced)

- Jack Ryan (January 21, 1975 – August 24, 1976) (divorced)
- Michael O'Hara (August 27, 1976–1983) (divorced)
- Felipe de Alba (April 13, 1983 – April 14, 1983) (annulled)
- Frédéric Prinz von Anhalt (August 14, 1986 – present)

Gabor's high number of divorces inspired her to make numerous quotable puns and innuendos about her marital (and extramarital) history:

- "I am a marvelous housekeeper: Every time I leave a man I keep his house."
- Asked "How many husbands have you had?", she replied "You mean other than [or 'apart from'] my own?".

While Gabor was still married to Conrad Hilton, she once admitted to having sexual relations with her stepson Nicky, future husband of Elizabeth Taylor. In 1974, she purchased a home in Bel Air, Los Angeles, California from Elvis Presley's estate. It was originally built by Howard Hughes and featured a unique looking French style roof. Gabor's only child, a daughter named Constance Francesca Hilton, was born on March 10, 1947. According to Gabor's 1991 autobiography *One Lifetime Is Not Enough*, her pregnancy resulted from rape by then-husband Conrad Hilton. She was the only Gabor sister to ever become pregnant. In 2005, Gabor accused her daughter of larceny and fraud, alleging that she had forged her signature to get a $2 million loan on her mother's Bel Air house, and filed a lawsuit against Francesca in a California court. However, the Santa Monica Superior Court threw out the case due to Gabor's refusal to appear in court or to sign an affidavit that she indeed was a co-plaintiff on the original lawsuit filed by her husband, Frédéric Prinz von Anhalt. Gabor said in a November 27, 1991, interview with David Letterman that she is a Democrat.

Filmography

- *Lovely to Look At* (LeRoy, 1952)
- *We're Not Married* (Goulding, 1952)
- *Moulin Rouge* (Huston, 1952)

- *The Million Dollar Nickel* (1952) (short subject)
- *The Story of Three Loves* (Minnelli, 1953)
- *Lili* (Walters, 1953)
- *L'ennemi public no.1* ("The Most Wanted Man") (Verneuil, 1953)
- *Sangre y luces* ("Love in a Hot Climate") (Rouquier/Suey, 1954)
- *Ball der Nationen* ("Ball of the Nations") (Ritter, 1954)
- *3 Ring Circus* (Pevney, 1954)
- *Death of a Scoundrel* (Martin, 1956)
- *The Girl in the Kremlin* (Birdwell, 1957)
- *The Man Who Wouldn't Talk* (Wilcox, 1958)
- *Country Music Holiday* (Ganzer, 1958)
- *Touch of Evil* (Welles, 1958) (as a "guest star")
- *Queen of Outer Space* (Bernds, 1958)

Television

- *Jukebox Jury*, as musical judge (1953)
- *The Red Skelton Show* (1955), as Movie Star
- *Climax!* (1955), as Mme. Florizel, Princess Stephanie
- *The Milton Berle Show* (1956)
- *Sneak Preview* (1956)
- *The Ford Television Theatre* (1956), as Dara Szabo
- *The Ford Show, Starring Tennessee Ernie Ford* (October 18, 1956), as Herself
- *General Electric Theater* (1956–1961), as Gloria
- *Matinee Theatre* (1956–1958), as Eugenia
- *The Life of Riley* (1957), as Gigi
- *Playhouse 90* (1957), as Erika Segnitz, Marta Lorenz
- *The Pat Boone Chevy Showroom*, as Herself
- *Shower of Stars* (1958)
- *Lux Playhouse* (1959), as Helen
- *Queen of Outer Space* (1959), with Eric Fleming
- *Ninotchka* (1960)

Plays

Gabor appeared in several plays, most notably *Forty Carats*, on Broadway, and *Blithe Spirit* (as Elvira), in the national tour.

41

Bette Davis

Ruth Elizabeth "Bette" Davis (April 5, 1908 – October 6, 1989) was an American actress of film, television and theater. Noted for her willingness to play unsympathetic characters, she was highly regarded for her performances in a range of film genres; from contemporary crime melodramas to historical and period films and occasional comedies, though her greatest successes were her roles in romantic dramas. After appearing in Broadway plays, Davis moved to Hollywood in 1930, but her early films for Universal Studios were unsuccessful. She joined Warner Bros. in 1932 and established her career with several critically acclaimed performances. In 1937, she attempted to free herself from her contract and although she lost a well-publicized legal case, it marked the beginning of the most successful period of her career. Until the late 1940s, she was one of American cinema's most celebrated leading ladies, known for her forceful and intense style. Davis gained a reputation as a perfectionist who could be highly combative, and confrontations with studio executives, film directors and costars were often reported. Her forthright manner, clipped vocal style and ubiquitous cigarette contributed to a public persona which has often been imitated and satirized.

Davis was the co-founder of the Hollywood Canteen, and was the first female president of the Academy of Motion Picture Arts and Sciences. She won the Academy Award for Best Actress twice, was the first person to accrue 10 Academy Award nominations for acting, and was the first woman to receive a Lifetime Achievement Award from the American Film Institute. Her career went through several periods of eclipse, and she admitted that her success had often been at the expense of her personal relationships. Married four times, she was once widowed and thrice divorced, and raised her children as a single parent. Her final years were marred by a long period of ill health, but she continued acting until shortly before her death from breast cancer, with more than 100 films, television and theater roles to her credit. In 1999, Davis was placed second, after Katharine Hepburn, on the American Film Institute's list of the greatest female stars of all time.

Life and Career

Background and Early Acting Career

Ruth Elizabeth Davis, known from early childhood as "Betty"., was born in Lowell, Massachusetts, the daughter of Ruth ("Ruthie") Augusta (née Favor) and Harlow Morrell Davis, a patent attorney; her sister Barbara ("Bobby") was born October 25, 1909. The family was Protestant, of English, French, and Welsh ancestry. In 1915, Davis's parents separated and Betty and Bobby attended a Spartan boarding school called Crestalban in Lanesborough, which is located in the Berkshires. In 1921, Ruth Davis moved to New York City with her daughters, where she worked as a portrait photographer. Betty was inspired to become an actress after seeing Rudolph Valentino in *The Four Horsemen of the Apocalypse* (1921) and Mary Pickford in *Little Lord Fauntleroy* (1921), and changed the spelling of her name to "Bette" after Honoré de Balzac's *La Cousine Bette*. She received encouragement from her mother, who had aspired to become an actress. She attended Cushing Academy, a boarding school in Ashburnham, Massachusetts, where she met her future husband, Harmon O. Nelson, known as "Ham".

In 1926, she saw a production of Henrik Ibsen's *The Wild Duck* with Blanche Yurka and Peg Entwistle. Davis later recalled that it inspired her full commitment to her chosen career, and said, "Before that performance I *wanted* to be an actress. When it ended, I *had* to be an actress... exactly like Peg Entwistle.". She auditioned for admission to Eva LeGallienne's Manhattan Civic Repertory, but was rejected by LeGallienne who described her attitude as "insincere" and "frivolous". She was accepted by the John Murray Anderson School of Theatre, and studied dance with Martha Graham. She auditioned for George Cukor's stock theater company, and although he was not very impressed, he gave Davis her first paid acting assignment anyway – a one-week stint playing the part of a chorus girl in the play *Broadway*. She was later chosen to play Hedwig, the character she had seen Entwistle play, in *The Wild Duck*. After performing in Philadelphia, Washington and Boston, she made her Broadway debut in 1929 in *Broken Dishes*, and followed it with *Solid South*. A Universal Studios talent scout saw her perform and invited her to Hollywood for a screen test.

42

Helena Rubinstein

Helena Rubinstein (December 25, 1870 – April 1, 1965), a Polish cosmetics industrialist, founder and eponym of Helena Rubinstein, Incorporated, which made her one of the world's richest women.

Early Life

Rubinstein was born Chaja Rubinstein, the eldest of eight children, to Augusta Gitte (Gitel) Scheindel Silberfeld Rubinstein and Naftali Herz Horace Rubinstein; he was a shopkeeper in Kraków. For a short time, she studied medicine in Switzerland.

Move to Australia

She arrived in Australia in 1902, with no money and little English. Her stylish clothes and milky complexion did not pass unnoticed among the town's ladies, however, and she soon found enthusiastic buyers for the jars of beauty cream in her luggage. Spotting a market, she began to make her own. Fortunately, a key ingredient was readily at hand. Coleraine, in Western Victoria, where her uncle was a shopkeeper, might have been an "awful place" but it did not lack for lanolin. Sheep, some 75 million of them, were the wealth of the nation and the Western District's vast mobs of merinos produced the finest wool in the land, secreting abundant grease in the process. To disguise the sheep oil's pungent pong, Rubinstein experimented with lavender, pine bark and water lilies. She also managed to fall out with her uncle. After a stint as a bush governess, she got a job as a waitress at the Winter Garden tearooms in Melbourne. There, she found an admirer willing to stump up the funds to launch her Crème Valaze, supposedly including herbs imported "from the Carpathian Mountains". Costing ten pence and selling for six shillings, it walked off the shelves as fast as she could pack it in pots. Now calling herself Helena, Rubinstein could soon afford to open a salon in fashionable Collins Street, selling glamour as a science to clients whose skin was "diagnosed" and a suitable treatment "prescribed". Sydney was next, and within five years Australian operations were profitable enough to finance a Salon de Beauté Valaze in London. As such, Rubinstein formed one of the world's first cosmetic companies. Her business enterprise proved immensely successful and later in life she used her enormous wealth to support charitable institutions in the fields of education, art and health. Diminutive at 4 ft. 10 in. (147 cm), she rapidly expanded her operation. In 1908, her sister Ceska assumed the Melbourne shop's operation, when, with $100,000, Rubinstein moved to London and began what was to become an international enterprise. (Women at this time could not obtain bank loans, so the money was her own.)

Marriage and Children

In 1908, she married American journalist Edward William Titus in London. They had two sons, Roy Valentine Titus (London, December 12, 1909–New York, June 18, 1989) and Horace Titus (London, April 23, 1912–New York, May 18, 1958). They eventually moved to Paris where she opened a salon in 1912. Her husband helped with writing the publicity and set up a small publishing house, published *Lady Chatterley's Lover* and hired Samuel Putnam to translate famous model Kiki's memoirs. Rubinstein threw lavish dinner parties and became known for apocryphal quips, such as when an intoxicated French ambassador expressed vitriol toward Edith Sitwell and her brother Sacheverell: "Vos ancêtres ont brûlé Jeanne d'Arc!" Rubinstein, who knew little French, asked a guest what the ambassador had said. "He said, 'Your ancestors burned Joan of Arc.' " Rubinstein replied, "Well, someone had to do it." At another *fête*, Marcel Proust asked her what makeup a duchess might wear. She summarily dismissed him because "he smelt of mothballs." Rubinstein recollected later, "How was I to know he was going to be famous?"

Move to the United States

At the outbreak of World War I, she and Titus moved to New York City, where she opened a cosmetics salon in 1915, the forerunner of a chain throughout the country. This was the beginning of her vicious rivalry with the other great lady of the cosmetics industry, Elizabeth Arden. Both Rubinstein and Arden, who died within 18 months of each other, were social climbers. And they were both keenly aware of effective marketing and luxurious packaging, the attraction of beauticians in neat uniforms, the value of celebrity endorsements, the perceived value of overpricing and the promotion of the pseudo-science of skincare. From 1917, Rubinstein took on the manufacturing and wholesale distribution of her products. The "Day of Beauty" in the various salons became a great success. The purported portrait of Rubinstein in her advertising was of a middle-age mannequin with a gentile appearance. In 1928, she sold the American business to Lehman Brothers for $7.3 million, ($88 million in 2007). After the arrival of the Great Depression, she bought back the nearly worthless stock for less than $1 million and eventually turned the shares into values of multimillion dollars, establishing salons and outlets in almost a dozen U.S. cities. Her subsequent spa at 715 Fifth Avenue included a restaurant, a gymnasium and rugs by painter Joan Miró. She commissioned Salvador Dalí to design a powder compact as well a portrait of herself.

Divorce and Remarriage

In 1937, Rubinstein divorced Titus after a contentious marriage marked by his infidelities. Freed of her former marriage vows, in 1938 Helena readily married Prince Artchil Gourielli-Tchkonia (1895–1955), whose somewhat clouded materlineal claim to Georgian nobility, as that of Prince Artchil Gourielli-Tchkonia (sometimes spelled Courielli-Tchkonia; born at Georgia, 18 February 1895, died at New York City 21 November 1955), stemmed from his having been born a member of the untitled noble Tchkonia family of Guria, enticing the ambitious young man to appropriate the genuine title of his grandmother, born Princess Gourielli. Self-styled Prince Artchil Gourielli-Tchkonia, was 23-years younger than Rubinstein.

Eager for a regal title to call her own, Rubinstein pursued the handsome youth avidly; coming to name a male cosmetics line after her youthful prized catch. Some have claimed that the marriage was a marketing ploy, including Rubinstein's being able to pass herself off as Helena Princess Gourielli. A multimillionaire of contrasts, Rubinstein took a bag lunch to work and was very frugal in many matters but bought top-fashion clothing and valuable fine art and furniture. Concerning art, she founded the respectable Helena Rubinstein Pavilion of Contemporary Art in Tel Aviv and in 1957 she established the Helena Rubinstein travelling art scholarship in Australia. In 1953, she established the philanthropic Helena Rubinstein Foundation to provide funds to organizations specializing in health, medical research and rehabilitation as well as to the America Israel Cultural Foundation and scholarships to Israelis. In 1959, Rubinstein represented the U.S. cosmetics industry at the American National Exhibition in Moscow. A £300 annual Rubinstein Prize was awarded for portraits by Australian artsts from 1960. Prizewinners included Charles Blackman 1960; William Boissevain 1961; Margaret Olley 1962; Vladas Meskenas 1963; Judy Cassab 1964, 1965; Jack Carington Smith 1966. Death and Legacy

Mme. Rubinstein died April 1, 1965 and was buried in Mount Olivet Cemetery in Queens. Some of her estate, including African and fine art, Lucite furniture, and overwrought Victorian furniture upholstered in purple, was auctioned in 1966 at the Park-Bernet Galleries in New York. One of Rubinstein's numerous mantras was: "There are no ugly women, only lazy ones." A scholarly study of her exclusive beauty salons and how they blurred and influenced the conceptual boundaries at the time among fashion, art galleries, the domestic interior and versions of modernism is explored by Marie J. Clifford (*Winterthur Portfolio*, vol. 38). A feature-length documentary film, *The Powder and the Glory* (2009) by Ann Carol Grossman and Arnie Reisman, details the rivalry between Rubinstein and Elizabeth Arden.

43

Helen Keller

Helen Adams Keller (June 27, 1880 – June 1, 1968) was an American author, political activist, and lecturer. She was the first deafblind person to earn a Bachelor of Arts degree. The story of how Keller's teacher, Anne Sullivan, broke through the isolation imposed by a near complete lack of language, allowing the girl to blossom as she learned to communicate, has become widely known through the dramatic depictions of the play and film *The Miracle Worker*. A prolific author, Keller was well-traveled, and was outspoken in her opposition to war. A member of the Socialist Party of America and the Wobblies, she campaigned for women's suffrage, workers' rights, and socialism, as well as many other leftist causes.

Early Childhood and Illness

Helen Adams Keller was born on June 27, 1880, in Tuscumbia, Alabama. Her family lived on a homestead, Ivy Green, that Helen's grandfather had built decades earlier. Helen's father, Arthur H. Keller, spent many years as an editor for the Tuscumbia *North Alabamian* and had served as a captain for the Confederate Army. Helen's paternal grandmother was the second cousin of Robert E. Lee. Helen's mother, Kate Adams, was the daughter of Charles Adams. Though originally from Massachusetts, Charles Adams also fought for the Confederate Army during the American Civil War, earning the rank of brigadier-general. Helen's father's lineage can be traced to Casper Keller, a native of Switzerland. Coincidentally, one of Helen's Swiss ancestors was the first teacher for the deaf in Zurich. Helen reflects upon this coincidence in her first autobiography, stating "that there is no king who has not had a slave among his ancestors, and no slave who has not had a king among his." Helen Keller was not born blind and deaf; it was not until she was 19 months old that she contracted an illness described by doctors as "an acute congestion of the stomach and the brain", which might have been scarlet fever or meningitis. The illness did not last for a particularly long time, but it left her deaf and blind. At that time, she was able to communicate somewhat with Martha Washington, the six-year-old daughter of the family cook, who understood her signs; by the age of seven, she had over 60 home signs to communicate with her family. In 1886, her mother, inspired by an account in Charles Dickens' *American Notes* of the successful education of another deaf and blind woman, Laura Bridgman, dispatched young Helen, accompanied by her father, to seek out Dr. J. Julian Chisolm, an eye, ear, nose, and throat specialist in Baltimore, for advice. He subsequently put them in touch with Alexander Graham Bell, who was working with deaf children at the time. Bell advised the couple to contact the Perkins Institute for the Blind, the school where Bridgman had been educated, which was then located in South Boston.

Formal Education

Starting in May, 1888, Keller attended the Perkins Institute for the Blind. In 1894, Helen Keller and Anne Sullivan moved to New York to attend the Wright-Humason School for the Deaf, and to learn from Sarah Fuller at the Horace Mann School for the Deaf. In 1896, they returned to Massachusetts and Keller entered The Cambridge School for Young Ladies before gaining admittance, in 1900, to Radcliffe College, where she lived in Briggs Hall, South House. Her admirer, Mark Twain, had introduced her to Standard Oil magnate Henry Huttleston Rogers, who, with his wife, paid for her education. In 1904, at the age of 24, Keller graduated from Radcliffe, becoming the first deaf blind person to earn a Bachelor of Arts degree. She maintained a correspondence with the Austrian philosopher and pedagogue Wilhelm Jerusalem, who was one of the first to discover her literary talent.

Political Activities

Keller went on to become a world-famous speaker and

author. She is remembered as an advocate for people with disabilities, amid numerous other causes. She was a suffragist, a pacifist, an opponent of Woodrow Wilson, a radical socialist and a birth control supporter. In 1915 she and George Kessler founded the Helen Keller International (HKI) organization. This organization is devoted to research in vision, health and nutrition. In 1920 she helped to found the American Civil Liberties Union (ACLU). Keller traveled to over 39 countries with Sullivan, making several trips to Japan and becoming a favorite of the Japanese people. Keller met every U.S. President from Grover Cleveland to Lyndon B. Johnson and was friends with many famous figures, including Alexander Graham Bell, Charlie Chaplin and Mark Twain. Keller and Mark Twain were both considered radicals at the beginning of the 20th century, and as a consequence, their political views have been forgotten or glossed over in popular perception. Keller was a member of the Socialist Party and actively campaigned and wrote in support of the working class from 1909 to 1921. She supported Socialist Party candidate Eugene V. Debs in each of his campaigns for the presidency. Newspaper columnists who had praised her courage and intelligence before she expressed her socialist views now called attention to her disabilities.

Writings

Keller wrote a total of 12 published books and several articles. One of her earliest pieces of writing, at age 11, was *The Frost King* (1891). There were allegations that this story had been plagiarized from *The Frost Fairies* by Margaret Canby. An investigation into the matter revealed that Keller may have experienced a case of cryptomnesia, which was that she had Canby's story read to her but forgot about it, while the memory remained in her subconscious. At age 22, Keller published her autobiography, *The Story of My Life* (1903), with help from Sullivan and Sullivan's husband, John Macy. It includes words that Keller wrote and the story of her life up to age 21, and was written during her time in college. Keller wrote *The World I Live In* in 1908 giving readers an insight into how she felt about the world. *Out of the Dark*, a series of essays on socialism, was published in 1913. When Keller was young, Anne Sullivan introduced her to Phillips Brooks, who introduced her to Christianity, Keller famously saying: "I always knew He was there, but I didn't know His name! Her spiritual autobiography, *My Religion*, was published in 1927 and then in 1994 extensively revised and re-issued under the title *Light in My Darkness*. It advocates the teachings of Emanuel Swedenborg, the Christian revelator and theologian who gives a spiritual interpretation of the teachings of the Bible and who claims that the second coming of Jesus Christ has already taken place. Adherents use several names to describe themselves, including Second Advent Christian, Swedenborgian and New Church.

Later Life

Keller suffered a series of strokes in 1961 and spent the last years of her life at her home. On September 14, 1964, President Lyndon B. Johnson awarded her the Presidential Medal of Freedom, one of the United States' highest two civilian honors. In 1965 she was elected to the National Women's Hall of Fame at the New York World's Fair. Keller devoted much of her later life to raising funds for the American Foundation for the Blind. She died in her sleep on June 1, 1968, at her home, Arcan Ridge, located in Easton, Connecticut. A service was held in her honor at the National Cathedral in Washington, D.C., and her ashes were placed there next to her constant companions, Anne Sullivan and Polly Thompson.

Portrayals

Keller's life has been interpreted many times. She appeared in a silent film, *Deliverance* (1919), which told her story in a melodramatic, allegorical style. She was also the subject of the documentaries *Helen Keller in Her Story*, narrated by Katharine Cornell, and *The Story of Helen Keller*, part of the Famous Americans series produced by Hearst Entertainment. *The Miracle Worker* is a cycle of dramatic works ultimately derived from her autobiography, *The Story of My Life*. The various dramas each describe the relationship between Keller and Sullivan, depicting how the teacher led her from a state of almost feral wildness into education, activism, and intellectual celebrity.

44

Coco Chanel

Gabrielle Bonheur "Coco" Chanel (19 August 1883 – 10 January 1971)

> " Fashion is not something that exists in dresses only. Fashion is in the sky, in the street, fashion has to do with ideas, the way we live, what is happening. "
>
> —*Coco Chanel*

Coco Chanel was a leading French modernist designer, whose patterns of simplicity and style revolutionised women's clothing. She was the only designer to be listed in Time 100 most influential people of the Twentieth Century. In the 1960s, a broadway musical was made about her life starring Katharine Hepburn. It was in the 1920s, that Coco Chanel left a lasting mark on Women's fashion and design. Up until the First World War, women's clothing had been quite restrictive and tended to involve full length skirts which were impractical for many activities. Coco Chanel helped create women's clothing that was simpler and more practical. She also introduced trousers and suits for women - something which had not been done before.

> "Fashion has become a joke. The designers have forgotten that there are women inside the dresses. Most women dress for men and want to be admired. But they must also be able to move, to get into a car without bursting their seams! Clothes must have a natural shape."
>
> —*Coco Chanel*

She also created her famous Chanel No.5 scent and this has been a lasting trademark. Most sources suggest she was born in 1883, though this was a closely guarded fact - with Coco not keen on revealing her birth date. Orphaned from an early age, she worked with her sister in a milliner in Deauville. Later, she opened a shop in 1912 and after a spell of nursing during the first world war founded a couture house in the Rue Cambon in Paris.

> It was in the post war period that she felt the need for a revolution in women's clothes. She began by liberating women from the bondage of the corset and encouraged a casual but elegant range of clothes. With a black sweater and 10 rows of pearls Chanel revolutionized fashion'
>
> —*Dior on Coco Chanel*

The 1920s were a significant period of liberation for women. It was a decade where women received the vote in several western countries. It was also a time, when women were increasingly seen in professions and jobs, previously the reserve of men. Her fashion symbolised some of these social and political changes. Significant items of clothing Coco Chanel helped pioneer included:

- the collarless cardigan jacket
- the bias cut dress - labelled a Ford by one critic because everyone had one.
- The shoe string shoulder strap.
- The floating evening scarf
- The wearing together of junk and real jewels.

In 1938 she retired from the fashion business. However, 16 years later she made a determined comeback after becoming fed up at seeing French fashion become dominated by men. Her first post war collection was not well received by the critics but it proved immensely popular with the general public. Rich and famous women once again adopted the Chanel look and she had shown her lasting influence on the industry. She prided herself on her great taste, fashion and practicality combined with an awareness of what people wanted. It was this that made her the most recognisable name in world fashion.

45

Emily Bronte

Emily Bronte (July 30, 1818 — December 19, 1848) - Poet and Novelist; famous for her classic novel *Wuthering Heights*.

> With wide-embracing love
> Thy Spirit animates eternal years,
> Pervades and broods above,
> Changes, sustains, dissolves, creates, and rears.
> Emily Bronte *No Coward Soul Is Mine* (1848)

Emily Bronte was born 30 July 1818 in Thornton, Near Bradford in Yorkshire. She was the fifth of six children, including Anne and Charlotte Bronte, who both became writers as well. When Emily was six years old, the Bronte family moved to the village of Haworth, a village nestled in the windswept moors of West Yorkshire, which later inspired many of her writings:

> Photo left - portrait by her brother.
> A heaven so clear, an earth so calm,
> So sweet, so soft, so hushed an air;
> And, deepening still the dreamlike charm,
> Wild moor-sheep feeding everywhere.
> Emily Bronter, *A Little While, a Little While* (1846) Stanza vii.

Her father was made the local curate of Haworth and the family lived there for the remainder of their lives. The old vicarage is now a museum dedicated to the Brontes. Shortly after moving to Howarth, Emily's mother passed away. The girls were then sent to the Clergy Daughters' school at Cowen Bridge. In the aftermath of her mother's passing, this was a traumatic experience as the sisters found the school harsh and unsympathetic. This school experience was incorporated into Charlotte Bronte's *Jane Eyre*. During a typhus epidemic, Emily lost two of her sisters (Maria and Elizabeth) to the illness; shortly afterwards Emily returned home where she was educated by her father and aunt. For a brief period, when she was 17, Emily went to Roe Head girl's school where Charlotte was a teacher. But, due to homesickness she soon returned home. The sisters hoped one day to set up their own school, though this never materialised. But, to gain experience, Emily became a teacher in Halifax in September 1838. However, she struggled to cope with the exhausting hours, and after a few months returned to Haworth. Apart from a brief stay in a girls academy in Belgium, Emily spent most of her later life in Haworth, where she concentrated on domestic tasks looking after her brother and family. Like her father she seems to have preferred a quiet, reclusive life. As a character in her novel writes:

> "I'm now quite cured of seeking pleasure in society, be it country or town. A sensible man ought to find sufficient company in himself."
>
> *Mr. Lockwood (Ch. III) - Wuthering Heights (1847)*

This certainly applied to her father, who was quite reclusive and liked to dine alone in his room. Domestic life for Emily, was undoubtedly made difficult by her brother, Branwell who suffered from mood swings, influenced by his alcohol and drug addictions. Branwell died in 1848, shortly before Emily. From an early age, Emily began writing displaying a vivid imagination. Her early writings were in collaboration with her sisters and brothers about an imaginary world (Gondal saga). Only small fragments remain from this period. She continued writing throughout her life, though it became an increasingly private affair; initially she disliked the idea of her poems being published though she was persuaded on finding her sisters had been writing similar poems. In 1846, the three Bronte sisters published a collection of poems under the pseudonyms Currer Bell (Charlotte), Ellis Bell (Emily) and Acton Bell (Anne). The fact they chose masculine names suggests they wanted to avoid the prejudgment of female writers. At the time, it was rare for women writers to be published.

In 1847, she published her only novel *Wuthering Heights*. Based on the windswept moors of Haworth, it is a powerful tale of love, hate, sorrow and death; it later became a classic of English literature. Though at the time, its innovative structure and complexity led to mixed reviews. In 1850, her sister Charlotte republished the book under Emily's real name. Frail throughout her life, Emily fell seriously sick in the autumn of 1848. Her health was undoubtedly harmed by unsanitary water which drained from the nearby churchyard. Shortly after her brother's funeral she caught a serious cold, and refusing medical help, she died on 19 December 1848. Emily Bronte left little writings about herself. Often, her poems and novel have been scrutinised for autobiographical hints. However, it is difficult to fully ascertain which poems are just imagination and which relate to part of her character. Her writings have been included in the braod period of romanticism. They range from stark reminders of the harshness of life, to the potential beauty and power of love and a mystical power of nature.

'Twas grief enough to think mankind
All hollow servile insincere
But worse to trust to my own mind
And find the same corruption there
Emily Bronte - *I Am the Only Being* (1836)
Then dawns the Invisible; the Unseen its truth reveals;
My outward sense is gone, my inward essence feels —
Its wings are almost free, its home, its harbour found;
Measuring the gulf, it stoops and dares the final bound

—Emily Bronte - The Prisoner (October 1845)

46

Amelia Earhart

Amelia Mary Earhart (born July 24, 1897; missing July 2, 1937, declared legally dead January 5, 1939) was a noted American aviation pioneer and author. Earhart was the first woman to receive the U.S. Distinguished Flying Cross, awarded for becoming the first aviatrix to fly solo across the Atlantic Ocean. She set many other records, wrote best-selling books about her flying experiences and was instrumental in the formation of The Ninety-Nines, an organization for female pilots. Earhart joined the faculty of the world-famous Purdue University aviation department in 1935 as a visiting faculty member to counsel women on careers and help inspire others with her love for aviation. She was also a member of the National Woman's Party, and an early supporter of the Equal Rights Amendment. During an attempt to make a circumnavigational flight of the globe in 1937 in a Purdue-funded Lockheed Model 10 Electra, Earhart disappeared over the central Pacific Ocean near Howland Island. Fascination with her life, career and disappearance continues to this day.

Early Life

Childhood

Amelia Mary Earhart, daughter of Samuel "Edwin" Stanton Earhart (March 28, 1867) and Amelia "Amy" Otis Earhart (1869–1962), was born in Atchison, Kansas, in the home of her maternal grandfather, Alfred Gideon Otis (1827–1912), a former federal judge, president of the Atchison Savings Bank and a leading citizen in Atchison. This was the second child in the marriage as an infant was stillborn in August 1896. Alfred Otis had not initially favored the marriage and was not satisfied with Edwin's progress as a lawyer. Earhart was named, according to family custom, after her two grandmothers (Amelia Josephine Harres and Mary Wells Patton). From an early age Earhart, nicknamed "Meeley" (sometimes "Millie") was the ringleader while younger sister (two years her junior), Grace Muriel Earhart (1899–1998), nicknamed "Pidge," acted the dutiful follower. Both girls continued to answer to their childhood nicknames well into adulthood. Their upbringing was unconventional since Amy Earhart did not believe in molding her children into "nice little girls." Meanwhile their maternal grandmother disapproved of the "bloomers" worn by Amy's children and although Earhart liked the freedom they provided, she was aware other girls in the neighborhood did not wear them.

Early Influence

A spirit of adventure seemed to abide in the Earhart children with the pair setting off daily to explore their neighborhood. As a child, Earhart spent long hours playing with Pidge, climbing trees, hunting rats with a rifle and "belly-slamming" her sled downhill. Although this love of the outdoors and "rough-and-tumble" play was common to many youngsters, some biographers have characterized the young Earhart as a tomboy. The girls kept "worms, moths, katydids and a tree toad" in a growing collection gathered in their outings. In 1904, with the help of her uncle, she cobbled together a home-made ramp fashioned after a roller coaster she had seen on a trip to St. Louis and secured the ramp to the roof of the family toolshed. Earhart's well-documented first flight ended dramatically. She emerged from the broken wooden box that had served as a sled with a bruised lip, torn dress and a "sensation of exhilaration." She exclaimed, "Oh, Pidge, it's just like flying!" Although there had been some missteps in his career up to that point, in 1907 Edwin Earhart's job as a claims officer for the Rock Island Railroad led to a transfer to Des Moines, Iowa. The next year, at the age of 10, Earhart saw her first aircraft at the Iowa State Fair in Des Moines. Her father tried to interest her and her sister in taking a flight. One look at the rickety old "flivver" was enough for Earhart, who promptly asked if they could go back to the merry-go-round. She later described the biplane as "a thing of rusty wire and wood and not at all interesting."

Education

The two sisters, Amelia and Muriel (she went by her middle name from her teens on), remained with their grandparents in Atchison, while their parents moved into new, smaller quarters in Des Moines. During this period, Earhart received a form of home-schooling together with her sister, from her mother and a governess. She later recounted that she was "exceedingly fond of reading" and spent countless hours in the large family library. In 1909, when the family was finally reunited in Des Moines, the Earhart children were enrolled in public school for the first time with Amelia Earhart entering the seventh grade at the age of 12 years.

Family Fortunes

While the family's finances seemingly improved with the acquisition of a new house and even the hiring of two servants, it soon became apparent Edwin was an alcoholic. Five years later (in 1914), he was forced to retire and although he attempted to rehabilitate himself through treatment, he was never reinstated at the Rock Island Railroad. At about this time, Earhart's grandmother Amelia Otis died suddenly, leaving a substantial estate that placed her daughter's share in trust, fearing that Edwin's drinking would drain the funds. The Otis house and all of its contents, was auctioned; Earhart was heart-broken and later described it as the end of her childhood. In 1915, after a long search, Earhart's father found work as a clerk at the Great Northern Railway in St. Paul, Minnesota, where Earhart entered Central High School as a junior. Edwin applied for a transfer to Springfield, Missouri, in 1915 but the current claims officer reconsidered his retirement and demanded his job back, leaving the elder Earhart with nowhere to go. Facing another calamitous move, Amy Earhart took her children to Chicago where they lived with friends. Earhart made an unusual condition in the choice of her next schooling; she canvassed nearby high schools in Chicago to find the best science program. She rejected the high school nearest her home when she complained that the chemistry lab was "just like a kitchen sink." She eventually was enrolled in Hyde

Park High School but spent a miserable semester where a yearbook caption captured the essence of her unhappiness, "A.E. – the girl in brown who walks alone."

Earhart graduated from Hyde Park High School in 1916. Throughout her troubled childhood, she had continued to aspire to a future career; she kept a scrapbook of newspaper clippings about successful women in predominantly male-oriented fields, including film direction and production, law, advertising, management and mechanical engineering. She began junior college at Ogontz School in Rydal, Pennsylvania but did not complete her program. During Christmas vacation in 1917, Earhart visited her sister in Toronto. World War I had been raging and Earhart saw the returning wounded soldiers. After receiving training as a nurse's aide from the Red Cross, she began work with the Volunteer Aid Detachment at Spadina Military Hospital. Her duties included preparing food in the kitchen for patients with special diets and handing out prescribed medication in the hospital's dispensary.

Early Flying Experiences

At about that time, with a young woman friend, Earhart visited an air fair held in conjunction with the Canadian National Exposition in Toronto. One of the highlights of the day was a flying exhibition put on by a World War I "ace." The pilot overhead spotted Earhart and her friend, who were watching from an isolated clearing and dived at them. "I am sure he said to himself, 'Watch me make them scamper,'" she said. Earhart stood her ground as the aircraft came close. "I did not understand it at the time," she said, "but I believe that little red airplane said something to me as it swished by." By 1919 Earhart prepared to enter Smith College but changed her mind and enrolled at Columbia University signing up for a course in medical studies among other programs. She quit a year later to be with her parents who had reunited in California.

Aviation Career and Marriage

According to the *Boston Globe*, Earhart was "one of the best women pilots in the United States," although this characterization has been disputed by aviation experts and experienced pilots in the decades since. She was an intelligent and competent pilot, but hardly a brilliant aviator, whose early efforts were characterized as inadequate by more seasoned flyers. One serious miscalculation occurred during a record attempt that had ended with her spinning down through a cloud bank, only to emerge at 3,000 ft (910 m). Experienced pilots admonished her, "Suppose the clouds had closed in until they touched the ground?" Earhart was chagrined, yet acknowledged her limitations as a pilot and continued to seek out assistance throughout her career from various instructors. By 1927, "Without any serious incident, she had accumulated nearly 500 hours of solo flying – a very respectable achievement."

1928 Transatlantic Flight

After Charles Lindbergh's solo flight across the Atlantic in 1927, Amy Phipps Guest, (1873–1959), expressed interest in being the first woman to fly (or be flown) across the Atlantic Ocean. After deciding the trip was too perilous for her to undertake, she offered to sponsor the project, suggesting they find "another girl with the right image." While at work one afternoon in April 1928, Earhart got a phone call from Capt. Hilton H. Railey, who asked her, "Would you like to fly the Atlantic?" The project coordinators (including book publisher and publicist George P. Putnam) interviewed Earhart and asked her to accompany pilot Wilmer Stultz and co-pilot/mechanic Louis Gordon on the flight, nominally as a passenger, but with the added duty of keeping the flight log. The team departed Trepassey Harbor, Newfoundland in a Fokker F.VIIb/3m on June 17, 1928, landing at Burry Port (near Llanelli), Wales, United Kingdom, exactly 20 hours and 40 minutes later. Since most of the flight was on "instruments" and Earhart had no training for this type of flying, she did not pilot the aircraft. When interviewed after landing, she said, "Stultz did all the flying—had to. I was just baggage, like a sack of potatoes." She added, "...maybe someday I'll try it alone." While in England, Earhart is reported as receiving a rousing welcome on June 19, 1928, when landing at Woolston in Southampton, England. She flew the Avro Avian 594 Avian III, SN: R3/AV/101 owned by Lady Mary Heath and later purchased the aircraft and had it shipped back to the United States (where it was assigned "unlicensed aircraft identification mark" 7083). When the Stultz, Gordon and Earhart flight crew returned to the United States, they were greeted with a ticker-tape parade in New York followed by a reception with President Calvin Coolidge at the White House.

Theories on Earhart's Disappearance

Many theories emerged after the disappearance of Earhart and Noonan. Two possibilities concerning the flyers' fate have prevailed among researchers and historians:

Crash and Sink Theory

Many researchers believe the Electra ran out of fuel and Earhart and Noonan ditched at sea. Navigator and aeronautical engineer Elgen Long and his wife Marie K. Long devoted 35 years of exhaustive research to the "crash and sink" theory, which is the most widely accepted explanation for the disappearance. Capt. Laurance F. Safford, USN, who was responsible for the interwar Mid Pacific Strategic Direction Finding Net, and the decoding of the Japanese PURPLE cipher messages for the attack on Pearl Harbor, began a lengthy analysis of the Earhart flight during the 1970s. His research included the intricate radio transmission documentation. Safford came to the conclusion, "poor planning, worse execution." Rear Admiral Richard R. Black, USN, who was in administrative charge of the Howland Island airstrip and was present in the radio room on the *Itasca,* asserted in 1982 that "the Electra went into the sea about 10 am, July 2, 1937 not far from Howland". British aviation historian Roy Nesbit interpreted evidence in contemporary accounts and Putnam's correspondence and concluded Earhart's Electra was not fully fueled at Lae. William L. Polhemous, the navigator on Ann Pellegreno's 1967 flight which followed Earhart and Noonan's original flight path, studied navigational tables for July 2, 1937 and thought Noonan may have miscalculated the "single line approach" intended to "hit" Howland.

David Jourdan, a former Navy submariner and ocean engineer specializing in deep-sea recoveries, has claimed any transmissions attributed to Gardner Island were false. Through his company Nauticos he extensively searched a 1,200-square-mile (3,100 km) quadrant north and west of Howland Island during two deep-sea sonar expeditions (2002 and 2006, total cost $4.5 million) and found nothing. The search locations were derived from the line of position (157–337) broadcast by Earhart on July 2, 1937. Nevertheless, Elgen Long's interpretations have led Jourdan to conclude, "The analysis of all the data we have – the fuel analysis, the radio calls, other things – tells me she went into the water off Howland." Earhart's stepson George Palmer Putnam Jr. has been quoted as saying he believes "the plane just ran out of gas." Susan Butler, author of the "definitive" Earhart biography *East to the Dawn*, says she thinks the aircraft went into the ocean out of sight of Howland Island and rests on the seafloor at a depth of 17,000 feet (5 km). Tom D. Crouch, Senior Curator of the National Air and Space Museum, has said the Earhart/Noonan Electra is "18,000 ft. down" and may even yield a range of artifacts that could rival the finds of the *Titanic*, adding, "...the mystery is part of what keeps us interested. In part, we remember her because she's our favorite missing person."

Legacy

Amelia Earhart was a widely known international celebrity during her lifetime. Her shyly charismatic appeal, independence, persistence, coolness under pressure, courage and goal-oriented career along with the circumstances of her disappearance at a young age have driven her lasting fame in popular culture. Hundreds of articles and scores of books have been written about her life which is often cited as a motivational tale, especially for girls. Earhart is generally regarded as a feminist icon. Earhart's accomplishments in aviation inspired a generation of female aviators, including the more than 1,000 women pilots of the Women Airforce Service Pilots (WASP) who ferried military aircraft, towed gliders, flew target practice aircraft, and served as transport pilots during World War II. A small section of Earhart's Lockheed Electra starboard engine nacelle recovered in the aftermath of the Hawaii crash has been confirmed as authentic and is now regarded as a control piece that will help to authenticate possible future discoveries. The evaluation of the scrap of metal was featured on an episode of *History Detectives* on Season 7 in 2009.

47

Raisa Gorbachova

Raisa Maximovna Gorbachova (5 January 1932 – 20 September 1999) was a major fundraiser for preservation of the Russian heritage, for new talents' education and for children's blood cancer treatment programs in Russia. She was the wife of Mikhail Gorbachev and was normally referred to as Raisa Gorbachev.

Biography

Raisa Titarenko was born in the city of Rubtsovsk in the Altai region of Siberia, the eldest of three children of Maxim Andreyevich Titarenko, a railway engineer originally from Chernihiv, Ukraine, and his Siberian wife, Alexandra Petrovna Porada, originally from Veseloyarsk. She spent her childhood years living in the Ural Mountains region, and met her future husband while studying philosophy in Moscow. She earned an advanced degree at the Moscow State Pedagogical Institute, and taught briefly at the Moscow State University. They married in September 1953 and moved to Mikhail's home region of Stavropol in southern Russia upon graduation. There, she taught Marxist-Leninist philosophy and defended her sociology research thesis about kolkhoz life. She gave birth to their only child, daughter Irina Mihailovna Virganskaya in 1957. When her husband returned to Moscow as a rising Soviet Communist Party official, Raisa Gorbachova took a post of a lecturer at her alma mater, Moscow State University. She left the post when Mikhail Gorbachev became a leader of the Soviet Union in 1985. Her public appearances beside her husband as First Lady were a novelty at home and went a long way in humanizing the country's image. Her dynamic personality and style caught the attention of Western media and observers. This contributed to mollifying the Western perception of the country as an "Evil Empire" that Ronald Reagan's anti-Soviet discourse had helped solidify.

Gorbachova made a $100,000 contribution to the charity "From hematologists of the world to children" when Prof. Rumiantsev and others addressed her in 1989. This and further donations raised by Gorbachyovs helped to buy equipment for blood banks and to train Russian doctors abroad. As a result, country-wide children's leukemia survival rates have since improved (Transcripts 2000). On 1 June 1990, Gorbachova accompanied U.S. First Lady Barbara Bush to Wellesley College in Massachusetts. Both women spoke before the graduating class during the commencement service, touching upon the role of women in modern society. Their addresses were covered on live television by all of the American broadcast networks. The CNN cable network provided live television coverage of their speeches around the world. The events of the Soviet Coup of 1991 left a scar on Gorbachova. The political turmoil that followed pushed aside Gorbachevs' life from the headlines. In 1997 she established Raisa Maksimovna's Club, meant to galvanize the participation of women in politics. Gorbachova was diagnosed with leukemia and died on 20 September 1999 at Münster University Hospital in Germany, aged 67. She is interred at the Novodevichy Cemetery in Moscow.

Books

- Gorbachyova, Raisa Maksimovna (1991). *Ya nadejus'....* Novosti.
- *Raisa. Vospominaniya, dnevniki, interview, statyi, telegrammy*. Moscow: Vagrius Petro-News. 2000.

48

Katharine Hepburn

Katharine Hepburn is one of the most famous actresses of the twentieth century. In a career lasting several decades she landed four Oscars - a record even today. She was an unconventional Hollywood actress, fiercely independent and often displaying a standoffish attitude to the media. However, her wide variety of roles and acting skills made her popular on screen and she was ranked the greatest female film star by the American Film Institute.

Early Life

Katharine Hepburn was born in Hartford, Connecticut. Her mother, Katharine Martha Haughton was a suffragette and her strong views and independence influenced the young Katharine. As a teenager, Katharine was free spirited getting involved in sports such as swimming, skating and gymnastics; she had a fearless streak and was suspended from school for smoking and breaking curfew. She later admitted to going swimming naked in the middle of the night. One incident which had a bearing on her early life. It involved finding her dear brother hanging from a rafter by a piece of rope. Her family tried to deny it was suicide, but it looked as if it was and the incident had a lasting impact on Katherine. Aged 21, Katherine married for the first time to socialite Ludlow Ogden Smith. The marriage did not last a long time and they divorced six years later. However, they remained friends and Katherine remained grateful for his support in her early years.

Early Acting Career

Katherine's early acting career was developed on stage and from this she graduated to film. By 1933 she had won her first Oscar for her performance in *Morning Glory*, a story about a woman who rejects romance for her work. There then followed a series of successful films such as:

- Alice Adams - earned her a second Oscar nomination
- State of the Union - directed by Frank Capra. Hepburn plays opposite Spencer Tracy in a film about an idealistic industrialists foray in to politics.

By the late 1930s, her acting career had started to decline, and she was once even labelled as 'toxic for the box office' along with people like Fred Astaire and Marlene Dietrich. A string of forgettable films was not helped by her dismissiveness towards other female actresses and reluctance to sign autographs or give interviews. She always retained a certain reluctance to give interviews to the press until her later life. She was nearly cast in the Oscar Winning film - *Gone with the Wind*. But, she wasn't keen and the role was giving to Vivien Leigh. After the war, Hepburn's career picked up. She gained an Oscar nomination in 1951 for her portrayal of a stern missionary in the film - *African Queen*. She starred alongside Humphrey Bogart. The film was a great success, though she ended up becoming quiet ill with dysentery and malaria due to the water. She later wrote about this difficult experience.

Notable Films

- Summertime (1955)
- The Rainmaker (1956)
- Guess Who's Coming to Dinner (1967) winner of 2 academy awards including Katherine Hepburn as best actress. Starred alongside Sidney Poitier and Spencer Tracy. It tells of a groundbreaking interracial marriage which was set against the backdrop of the civil rights movements of the 1960s
- The Lion in Winter. Katharine played Eleanor of Aquitaine.
- On Golden Pond (1981) - another Oscar winning performance by Katharine Hepburn

49

Simone de Beauvoir

Simone-Ernestine-Lucie-Marie Bertrand de Beauvoir, (January 9, 1908 – April 14, 1986), was a French existentialist philosopher, public intellectual, and social theorist. She wrote novels, essays, biographies, an autobiography in several volumes, and monographs on philosophy, politics, and social issues. She is now best known for her metaphysical novels, including *She Came to Stay* and *The Mandarins*, and for her 1949 treatise *The Second Sex*, a detailed analysis of women's oppression and a foundational tract of contemporary feminism. She is also noted for her lifelong polyamorous relationship with Jean-Paul Sartre.

Early Years

Simone de Beauvoir was born in Paris, the eldest daughter of Georges Bertrand de Beauvoir, a legal secretary who once aspired to be an actor, and Françoise (born) Brasseur, a wealthy banker's daughter and devout Catholic. Her younger sister, Hélène, was born two years later. The family struggled to maintain their bourgeois status after losing much of their fortune shortly after World War I, and Françoise insisted that the two daughters be sent to a prestigious convent school. Beauvoir herself was deeply religious as a child—at one point intending to become a nun—until a crisis of faith at age 14. She remained an atheist for the rest of her life. Beauvoir was intellectually precocious from a young age, fueled by her father's encouragement: he reportedly would boast, "Simone thinks like a man!" After passing baccalaureate exams in mathematics and philosophy in 1925, she studied mathematics at the Institut Catholique and literature/languages at the

Institut Sainte-Marie. She then studied philosophy at the Sorbonne, writing her thesis on Leibniz for Léon Brunschvicg. She first worked with Maurice Merleau-Ponty and Claude Lévi-Strauss when all three completed their practice teaching requirements at the same secondary school. Although not officially enrolled, she sat in on courses at the École Normale Supérieure in preparation for the *agrégation* in philosophy, a highly competitive postgraduate examination which serves as a national ranking of students. It was while studying for the *agrégation* that she met *École Normale* students Sartre, Paul Nizan, and René Maheu (who gave her the lasting nickname "Castor", or beaver). The jury for the *agrégation* narrowly awarded Sartre first place instead of Beauvoir, who placed second and, at age 21, was the youngest person ever to pass the exam.

Middle Years

Sartre

Sartre was dazzlingly intelligent and was just under 5 feet (1.5 m) tall. During October 1929, the two became a couple and Sartre asked her to marry him. One day while they were sitting on a bench outside the Louvre, he said, "Let's sign a two-year lease". Near the end of her life, Beauvoir said, "Marriage was impossible. I had no dowry." So they became an imaginary married couple. Beauvoir chose to never marry and did not set up a joint household with Sartre. She never had children. This gave her time to earn an advanced academic degree, to join political causes and to travel, write, teach, and to have (male and female - the latter often shared) lovers.

Ephebophilia Debate

A number of de Beauvoir's young female lovers were underage, and the nature of some of these relationships, some of which she instigated while working as a school teacher, has led to a biographical controversy and debate over whether de Beauvoir had inclinations towards ephebophilia. A former student, Bianca Lamblin, originally Bianca Bienenfeld, later wrote critically about her seduction by her teacher, Simone de Beauvoir, when she was a 17-year-old lycee student in her book, *Mémoires d'une jeune fille dérangée.* In 1941, de Beauvoir was suspended from her teaching job, due to an accusation that she had, in 1939, seduced her 17-year-old lycee pupil Nathalie Sorokine. De Beauvoir would, along with other French intellectuals, later petition for an abolition of all age of consent laws in France.

Existentialist Ethics

In 1944 Beauvoir wrote *Pyrrhus et Cinéas*, a discussion of an existentialist ethics, which inspired her to write more on the subject. This book, *Pour Une Morale de L'ambiguïté* (*The Ethics of Ambiguity*, 1947) is perhaps the most accessible entry into French existentialism. Its simplicity keeps it understandable, in contrast to the abstruse character of Sartre's *Being and Nothingness*. The ambiguity about which Beauvoir writes clears up some inconsistencies that many, Sartre included, have found in major existentialist works such as *Being and Nothingness*.

Les Temps Modernes

At the end of World War II, Beauvoir and Sartre edited *Les Temps Modernes*, a political journal Sartre founded along with Maurice Merleau-Ponty and others. Beauvoir used *Les Temps Modernes* to promote her own work and explore her ideas on a small scale before fashioning essays and books. Beauvoir remained an editor until her death.

Sexuality, Existentialist Feminism, and the Second Sex

Chapters of *Le deuxième sexe* (*The Second Sex*) were originally published in *Les Temps modernes*, in June 1949. The second volume came a few months after the first in France. It was very quickly published in America as *The Second Sex*, due to the quick translation by Howard Parshley, as prompted by Blanche Knopf, wife of publisher Alfred A. Knopf. Because Parshley had only a basic familiarity with the French language, and a minimal understanding of philosophy (he was a professor of biology at Smith College), much of Beauvoir's book was mistranslated or inappropriately cut, distorting her intended message. For years Knopf prevented the introduction of a more accurate retranslation of Beauvoir's work, declining all proposals despite the efforts of existentialist scholars. Only in 2009 was there a second translation, to mark the 60th anniversary of the original publication. Constance Borde and Sheila Malovany-Chevallier produced the first integral translation, reinstating a third of the original work. Beauvoir anticipated the sexually charged feminism of Erica Jong and Germaine Greer. Algren was outraged by the frank way Beauvoir later described her American sexual experiences in *The Mandarins* (dedicated to Algren, on whom the character Lewis Brogan was based) and in her autobiographies. He vented his outrage when reviewing American translations of her work. Much material bearing on this episode in Beauvoir's life, including her love letters to Algren, entered the public domain only after her death. In the chapter "Woman: Myth and Reality" of *The Second Sex*, Beauvoir argued that men had made women the "Other" in society by putting a false aura of "mystery" around them. She argued that men used this as an excuse not to understand women or their problems and not to help them, and that this stereotyping was always done in societies by the group higher in the hierarchy to the group lower in the hierarchy. She wrote that this also happened on the basis of other categories of identity, such as race, class, and religion. But she said that it was nowhere more true than with sex in which men stereotyped women and used it as an excuse to organize society into a patriarchy.

The Second Sex, published in French, sets out a feminist existentialism which prescribes a moral revolution. As an existentialist, Beauvoir believed that *existence precedes essence*; hence one is not born a woman, but becomes one. Her analysis focuses on the Hegelian concept of the Other. It is the (social) construction of Woman as the quintessential Other that Beauvoir identifies as fundamental to women's oppression. The capitalized 'O' in "other" indicates the wholly other. Beauvoir argued that women have historically been considered deviant, abnormal. She said that even Mary Wollstonecraft considered men to be the ideal toward which women should aspire. Beauvoir said that this attitude limited women's success by maintaining the perception that they were a deviation from the normal, and were always outsiders attempting to emulate "normality". She believed that for feminism to move forward, this assumption must be set aside. Beauvoir asserted that women are as capable of choice as men, and thus can choose to elevate themselves, moving beyond the 'immanence' to which they were previously resigned and reaching 'transcendence', a position in which one takes responsibility for oneself and the world, where one chooses one's freedom. A critical essay, "Le Malentendu du Deuxième Sexe", was written by Suzanne Lilar in 1969. A new translation of *The Second Sex* for the first time gives access to the entire text in English, as Irene Gammel writes in a *Globe and Mail* review: "The single most important advantage of this new translation is its completeness, combined with the translators' courage to transpose Beauvoir's existential language, thereby giving readers a sense of Beauvoir's channelling of Hegel, Marx and others."

Death, Honors and Legacy

Beauvoir died of pneumonia in Paris, aged 78. She is buried next to Sartre at the Cimetière du Montparnasse in Paris. Since her death, her reputation has grown. Especially in academia, she is considered the mother of post-1968 feminism. There has also been a growing awareness of her as a major French thinker and existentialist philosopher. Contemporary discussion analyzes the influences of Beauvoir and Sartre on one another. She is seen as having influenced Sartre's masterpiece, *Being and Nothingness*, while also having written much on philosophy that is independent of Sartrean existentialism. Some scholars have explored the influences of her earlier philosophical essays and treatises upon Sartre's later thought. She is studied by many respected academics both within and outside philosophy circles, including Margaret A. Simons and Sally Scholtz. Beauvoir's life has also inspired numerous biographies. In 2006, the city of Paris commissioned architect Dietmar Feichtinger to design a sophisticated footbridge across the Seine River. The bridge was named the Passerelle Simone-de-Beauvoir in her honor. It leads to the new Bibliothèque nationale de France.

50

Marlene Dietrich

Marlene Dietrich (27 December 1901 – 6 May 1992) was a German actress and singer. Dietrich remained popular throughout her long career by continually re-inventing herself. In 1920s Berlin, she acted on the stage and in silent films. Her performance as Lola-Lola in *The Blue Angel*, directed by Josef von Sternberg, brought her international fame and a contract with Paramount Pictures in the US. Hollywood films such as *Shanghai Express* and *Desire* capitalised on her glamour and exotic looks, cementing her stardom and making her one of the highest paid actresses of the era. Dietrich became a US citizen in 1937; during World War II, she was a high-profile frontline entertainer. Although she still made occasional films in the post-war years, Dietrich spent most of the 1950s to the 1970s touring the world as a successful show performer. In 1999 the American Film Institute named Dietrich the ninth greatest female star of all time.

Childhood

Dietrich was born Maria Magdalene Dietrich on 27 December 1901 in Schöneberg, a district of Berlin, Germany. She was the younger of two daughters (her sister Elisabeth being a year older) of Louis Erich Otto Dietrich and Wilhelmina Elisabeth Josephine Felsing, who married in December 1898. Dietrich's mother was from a well-to-do Berlin family who owned a clockmaking firm and her father was a police lieutenant. Her father died in 1911. His best friend, Eduard von Losch, an aristocrat first lieutenant in the Grenadiers courted Wilhelmina and eventually married her in 1916, but he died soon after as a result of injuries sustained during World War

I. Eduard von Losch never officially adopted the Dietrich children, hence Dietrich's surname was never von Losch, as is sometimes claimed. She was nicknamed "Lena" and "Lene" (pronounced Lay-neh) within the family. Around the age of 11, she contracted her two first names to form the then-novel name of "Marlene".

Early Career

Her earliest professional stage appearances were as a chorus girl on tour with Guido Thielscher's Girl-Kabarett, vaudeville-style entertainments, and in Rudolf Nelson revues in Berlin. In 1922, Dietrich auditioned unsuccessfully for theatrical director and impresario Max Reinhardt's drama academy; however, she soon found herself working in his theaters as a chorus girl and playing small roles in dramas, without attracting any special attention at first. She made her film debut playing a bit part in the 1922 film, *So sind die Männer*. She met her future husband, Rudolf Sieber, on the set of another film made that year, *Tragödie der Liebe*. Dietrich and Sieber were married in a civil ceremony in Berlin on 17 May 1923 Her only child, daughter Maria Elisabeth Sieber, was born on 13 December 1924. Dietrich continued to work on stage and in film both in Berlin and Vienna throughout the 1920s. On stage, she had roles of varying importance in Frank Wedekind's *Pandora's Box*, William Shakespeare's *The Taming of the Shrew* and *A Midsummer Night's Dream* as well as George Bernard Shaw's *Back to Methuselah* and *Misalliance*. It was in musicals and revues, such as *Broadway*, *Es Liegt in der Luft* and *Zwei Krawatten*, however, that she attracted the most attention. By the late 1920s, Dietrich was also playing sizable parts on screen, including *Café Elektric* (1927), *Ich küsse Ihre Hand, Madame* (1928) and *Das Schiff der verlorenen Menschen* (1929).Film Star

On the strength of *The Blue Angel's* international success, and with encouragement and promotion from von Sternberg, who was already established in Hollywood, Dietrich then moved to the U.S. on contract to Paramount Pictures. The studio sought to market Dietrich as a German answer to MGM's Swedish sensation, Greta Garbo. Her first American film, *Morocco*, directed by von Sternberg, earned Dietrich her only Oscar nomination. However, at the time she knew very little English, and so spoke her lines phonetically. Dietrich starred in six films directed by von Sternberg at Paramount between 1930 and 1935: *Morocco*, *Dishonored*, *Shanghai Express*, *Blonde Venus*, *The Scarlet Empress*, and *The Devil is a Woman.* In Hollywood, von Sternberg worked very effectively with Dietrich to create the image of a glamorous femme fatale. He encouraged her to lose weight and coached her intensively as an actress – she, in turn, was willing to trust him and follow his sometimes imperious direction in a way that a number of other performers resisted.

World War II

Dietrich was known to have strong political convictions and the mind to speak them. In interviews, Dietrich stated that she had been approached by representatives of the Nazi Party to return to Germany, but had turned them down flat. Dietrich, a staunch anti-Nazi, became an American citizen in 1939. In December 1941, the U.S. entered World War II, and Dietrich became one of the first celebrities to raise war bonds. She toured the US from January 1942 to September 1943 (appearing before 250 000 troops on the Pacific Coast leg of her tour alone) and it is said that she sold more war bonds than any other star. During two extended tours for the USO in 1944 and 1945, she performed for Allied troops on the front lines in Algeria, Italy, England and France and went into Germany with Generals James M. Gavin and George S. Patton. When asked why she had done this, in spite of the obvious danger of being within a few kilometres of German lines, she replied, "aus Anstand" — "out of decency".

Stage and Cabaret

From the early 1950s until the mid-1970s, Dietrich worked almost exclusively as a highly-paid cabaret artist, performing live in large theaters in major cities worldwide. In 1953, Dietrich was offered a then-substantial $30,000 per week to appear live at the Sahara Hotel on the Las Vegas Strip. The show was short, consisting only of a few songs associated with her. Her daringly sheer "nude dress" — a heavily beaded evening gown of silk soufflé, which gave the illusion of transparency — designed by Jean Louis, attracted a lot of publicity. This engagement was so successful that she was signed to appear at the Café de Paris in London the following year, and her Las Vegas contracts were also renewed. Dietrich employed Burt Bacharach as her musical arranger starting in the mid-1950s; together they refined her nightclub act into a more ambitious theatrical one-woman show with an expanded repertoire. Her repertoire included songs from her films as well as popular songs of the day. Bacharach's arrangements helped to disguise Dietrich's limited vocal range – she was a contralto – and allowed her to perform her songs to maximum dramatic effect; together, they recorded four albums and several singles between 1957 and 1964. She would often perform the first part of her show in one of her body-hugging dresses and a swansdown coat, and change to top hat and tails for the second half of the performance. This allowed her to sing songs usually associated with male singers, like "One For My Baby" and "I've Grown Accustomed to Her Face".

Final Years

Dietrich's show business career largely ended on 29 September 1975, when she broke her leg during a stage performance in Sydney, Australia. Her husband, Rudolf Sieber, died of cancer on 24 June 1976.

51

Rosa Parks

Rosa Louise McCauley Parks (February 4, 1913 – October 24, 2005) was an African American civil rights activist and seamstress whom the U.S. Congress dubbed the "Mother of the Modern-Day Civil Rights Movement". Parks is famous for her refusal on December 1, 1955 to obey bus driver James Blake's demand that she relinquish her seat to a white man. Her subsequent arrest and trial for this act of civil disobedience triggered the Montgomery Bus Boycott, one of the largest and most successful mass movements against racial segregation in history, and launched Martin Luther King, Jr., one of the organizers of the boycott, to the forefront of the civil rights movement. Her role in American history earned her an iconic status in American culture, and her actions have left an enduring legacy for civil rights movements around the world.

Montgomery Bus Boycott

After a day at work at Montgomery Fair department store, Parks boarded the Cleveland Avenue bus at around 6 p.m., Thursday, December 1, 1955, in downtown Montgomery. She paid her fare and sat in an empty seat in the first row of back seats reserved for blacks in the "colored" section, which was near the middle of the bus and directly behind the ten seats reserved for white passengers. Initially, she had not noticed that the bus driver was the same man, James F. Blake, who had left her in the rain in 1943. As the bus traveled along its regular route, all of the white-only seats in the bus filled up. The bus reached the third stop in front of the Empire Theater, and several white passengers boarded. In 1900, Montgomery had passed a city ordinance for the purpose of segregating passengers by race. Conductors were given the power to assign seats to accomplish that purpose; however, no passengers would be required to move or give up their seat and stand if the bus was crowded and no other seats were available. Over time and by custom, however, Montgomery bus drivers had adopted the practice of requiring black riders to move whenever there were no white only seats left. So, following standard practice, bus driver Blake noted that the front of the bus was filled with white passengers and there were two or three men standing, and thus moved the "colored" section sign behind Parks and demanded that four black people give up their seats in the middle section so that the white passengers could sit. Years later, in recalling the events of the day, Parks said, "When that white driver stepped back toward us, when he waved his hand and ordered us up and out of our seats, I felt a determination cover my body like a quilt on a winter night."

By Parks' account, Blake said, "Y'all better make it light on yourselves and let me have those seats." Three of them complied. Parks said, "The driver wanted us to stand up, the four of us. We didn't move at the beginning, but he says, 'Let me have these seats.' And the other three people moved, but I didn't." The black man sitting next to her gave up his seat. Parks moved, but toward the window seat; she did not get up to move to the newly repositioned colored section. Blake then said, "Why don't you stand up?" Parks responded, "I don't think I should have to stand up." Blake called the police to arrest Parks. When recalling the incident for Eyes on the Prize, a 1987 public television series on the Civil Rights Movement, Parks said, "When he saw me still sitting, he asked if I was going to stand up, and I said, 'No, I'm not.' And he said, 'Well, if you don't stand up, I'm going to have to call the police and have you arrested.' I said, 'You may do that.'" During a 1956 radio interview with Sydney Rogers in West Oakland several months after her arrest, when asked why she had decided not to vacate her bus seat, Parks said, "I would have to know for once and for all what rights I had as a human being and a citizen of Montgomery, Alabama."

When Parks refused to give up her seat, a police officer

arrested her. As the officer took her away, she recalled that she asked, "Why do you push us around?" The officer's response as she remembered it was, "I don't know, but the law's the law, and you're under arrest." She later said, "I only knew that, as I was being arrested, that it was the very last time that I would ever ride in humiliation of this kind." Parks was charged with a violation of Chapter 6, Section 11 segregation law of the Montgomery City code, even though she technically had not taken up a white-only seat—she had been in a colored section. E.D. Nixon and Clifford Durr bailed Parks out of jail the evening of December 1. That evening, Nixon conferred with Alabama State College professor Jo Ann Robinson about Parks' case. Robinson, a member of the Women's Political Council (WPC), stayed up all night mimeographing over 35,000 handbills announcing a bus boycott. The Women's Political Council was the first group to officially endorse the boycott.

On Sunday, December 4, 1955, plans for the Montgomery Bus Boycott were announced at black churches in the area, and a front-page article in The Montgomery Advertiser helped spread the word. At a church rally that night, attendees unanimously agreed to continue the boycott until they were treated with the level of courtesy they expected, until black drivers were hired, and until seating in the middle of the bus was handled on a first-come basis. Four days later, Parks was tried on charges of disorderly conduct and violating a local ordinance. The trial lasted 30 minutes.

On Monday, December 5, 1955, after the success of the one-day boycott, a group of 16 to 18 people gathered at the Mt. Zion AME Zion Church to discuss boycott strategies. The group agreed that a new organization was needed to lead the boycott effort if it were to continue. Rev. Ralph David Abernathy suggested the name "Montgomery Improvement Association" (MIA). The name was adopted, and the MIA was formed. Its members elected as their president a relative newcomer to Montgomery, a young and mostly unknown minister of Dexter Avenue Baptist Church, Dr. Martin Luther King, Jr. That Monday night, 50 leaders of the African American community gathered to discuss the proper actions to be taken in response to Parks' arrest. E.D. Nixon said, "My God, look what segregation has put in my hands!" Parks was the ideal plaintiff for a test case against city and state segregation laws. While the 15-year-old Claudette Colvin, unwed and pregnant, had been deemed unacceptable to be the center of a civil rights mobilization, King stated that, "Mrs. Parks, on the other hand, was regarded as one of the finest citizens of Montgomery— not one of the finest Negro citizens, but one of the finest citizens of Montgomery." Parks was securely married and employed, possessed a quiet and dignified demeanor, and was politically savvy.

The day of Parks' trial — Monday, December 5, 1955 — the WPC distributed the 35,000 leaflets. The handbill read, "We are...asking every Negro to stay off the buses Monday in protest of the arrest and trial . . . You can afford to stay out of school for one day. If you work, take a cab, or walk. But please, children and grown-ups, don't ride the bus at all on Monday. Please stay off the buses Monday." It rained that day, but the black community persevered in their boycott. Some rode in carpools, while others traveled in black-operated cabs that charged the same fare as the bus, 10 cents. Most of the remainder of the 40,000 black commuters walked, some as far as 20 miles. In the end, the boycott lasted for 382 days. Dozens of public buses stood idle for months, severely damaging the bus transit company's finances, until the law requiring segregation on public buses was lifted. Some segregationists retaliated with terrorism. Black churches were burned or dynamited. Martin Luther King's home was bombed in the early morning hours of January 30, 1956, and E.D. Nixon's home was also attacked. However, the black community's bus boycott marked one of the largest and most successful mass movements against racial segregation. It sparked many other protests, and it catapulted King to the forefront of the Civil Rights Movement.

Through her role in sparking the boycott, Rosa Parks played an important part in internationalizing the awareness of the plight of African Americans and the civil rights struggle. King wrote in his 1958 book Stride Toward Freedom that Parks' arrest was the precipitating factor, rather than the cause, of the protest: "The cause lay deep in the record of similar injustices.... Actually, no one can understand the action of Mrs. Parks unless he realizes that eventually the cup of endurance runs over, and the human personality cries out, 'I can take it no longer.'" The Montgomery bus boycott was also the inspiration for the bus boycott in the township of Alexandria, Eastern Cape of South Africa which was one of the key events in the radicalization of the black majority of that country under the leadership of the African National Congress.

Death and Funeral

Rosa Parks resided in Detroit until she died at the age of ninety-two on October 24, 2005, about 19:00 EDT, in her apartment on the east side of the city. She had been diagnosed the previous year with progressive dementia. City officials in Montgomery and Detroit announced on October 27 that the front seats of their city buses would be reserved with black ribbons in honour of Parks until her funeral. Parks' coffin was flown to Montgomery and taken in a horse-drawn hearse to the St. Paul African Methodist Episcopal (AME) church, where she lay in repose at the altar, dressed in the uniform of a church deaconess, on October 29. A memorial service was held there the following morning, and one of the speakers, Secretary of State Condoleezza Rice, said that if it had not been for Rosa Parks, she would probably have never become the Secretary of State. In the evening the casket was transported to Washington, D.C., and taken, aboard a bus similar to the one in which she made her protest, to lie in honor in the U.S. Capitol Rotunda. An estimated 50,000 people viewed the casket there, and the event was broadcast on television on October 31.

Parks' funeral service, seven hours long, was held on Wednesday, November 2, at the Greater Grace Temple Church. After the funeral service, an honor guard from the Michigan National Guard laid the U.S. flag over the casket and carried it to a horse-drawn hearse, which had been intended to carry it, in daylight, to the cemetery. Rosa was interred between her husband and mother at Detroit's Woodlawn Cemetery in the chapel's mausoleum.

52

Jiang Qing

Jiang Qing (20 March 1914 – 14 May 1991) was the pseudonym that was used by Chinese leader Mao Zedong's last wife and major Communist Party of China power figure. She went by the stage name Lan Ping during her acting career, and was known by various other names during her life. She married Mao in Yan'an in November 1938, and is sometimes referred to as Madame Mao in Western literature, serving as Communist China's first first lady. Jiang Qing was most well-known for playing a major role in the Cultural Revolution (1966–76) and for forming the radical political alliance known as the "Gang of Four". She was named the "Great Flag-carrier of the Proletarian Culture". Jiang Qing served as Mao's personal secretary in the 1940s and was head of the Film Section of the CPC Propaganda Department in the 1950s. In the early 1960s, she made a bid for power during the Cultural Revolution (1966–1976), which resulted in widespread chaos within the communist party. In 1966 she was appointed deputy director of the Central Cultural Revolution Group and claimed real power over Chinese politics for the first time. She became one of the masterminds of the Cultural Revolution, and along with three others, held absolute control over all of the national institutions.

Around the time of Chairman Mao's death, Jiang Qing and her proteges maintained control of many of China's power institutions, including a heavy hand in the media and propaganda. However, Jiang Qing's political success was limited. When Mao died in 1976, Jiang lost the support and justification for her political activities. She was arrested in October 1976 by Hua Guofeng and his allies, and was subsequently accused of being counter-revolutionary. Since then, Jiang Qing and Lin Biao have been branded by official historical documents in China as the "Lin Biao and Jiang Qing Counter-revolutionary Cliques", to which most of the blame for the damage and devastation caused by the Cultural Revolution was assigned. The assessments of western scholars have not been as uniformly critical. When Jiang Qing, along with the rest of the gang, was finally arrested, many, if not most, Chinese citizens rejoiced. Though initially sentenced to execution, her sentence was commuted to life imprisonment in 1983, however, and in May 1991 she was released for medical treatment. Before returning to prison, she committed suicide.

Early Life

Jiang Qing was born in Zhucheng, Shandong Province on March 20, 1914. Her birth name was LĐ Shûméng. She was the only child of Li Dewen, a carpenter, and his subsidiary wife, or concubine. Her father ran his own carpentry and cabinet making shop. After a violent argument between her parents, her mother left with the child to work as a domestic servant. Some accounts claim that Jiang's mother also worked as a prostitute. From July 1931 to April 1933, L- Yúnhè attended Qingdao University in Qingdao. She met Yu Qiwei, a biology student three years her senior, who was an underground member of the Communist Party Propaganda Department. By 1932, they had fallen in love and were living together. She joined the "Communist Cultural Front," a circle of artists, writers, and actors, and performed in *Put Down Your Whip*, a renowned popular play about a woman who escapes from the Japanese-occupied northeastern China and performs in the streets to survive. In February 1933, L- Yúnhè took the oath of the Chinese Communist Party with Yu Qiwei at her side, and she was appointed member of the Chinese Communist Party youth wing. Yu Qiwei was arrested in April the same year, and L- Yúnhè fled to her parents' home in Shanghai. When she arrived in Shanghai, the Yu family did not acknowledge her. She departed, and was soon back at the drama school in Jinan where she was warmly received. Through friendships she had previously established, she received an introduction to attend Shanghai University for the summer where she also taught some general literacy classes. In October, she rejoined the Communist Youth League, and at the same time, began participating in an amateur drama troupe. In September 1934, Jiang Qing was arrested and jailed for her political activities in Shanghai, but was released three months later, in December of the same year. She then traveled to Beijing where she reunited with Yu Qiwei who had just been released following his prison sentence, and the two began living together again.

Rise to Power

Entry into Chinese Politics

From the 1940s on, Mao and Jiang quarreled frequently. After the founding of the People's Republic of China in 1949, Jiang became the nation's first lady. She worked as Director of film in the Central Propaganda Department, and as a member of the Ministry of Culture steering committee for the film industry. An uproar in 1950 led the investigation of *The Life of Wu Xun*, a film about a 19th century beggar who raised money to educate the poor. Jiang supported criticism of the film for celebrating counter-revolutionary ideas.

Death of Mao Zedong

By September 5, 1976, Mao's condition turned critical. Upon being contacted by Hua Guofeng, Jiang Qing returned from her trip and spent only a few moments in the hospital's Building 202, where Mao was being treated. Later she returned to her own residence in the Spring Lotus Chamber. On the afternoon of September 7, Mao took a turn for the worse. Mao had just fallen asleep and needed to rest, but Jiang Qing insisted on rubbing his back and moving his limbs, and she sprinkled powder on his body. The medical team protested that the dust from the powder was not good for his lungs, but she instructed the nurses on duty to follow her example later. The next morning, September 8, she went again. This time she wanted the medical staff to change Mao's sleeping position, claiming that he had been lying too long on his left side. The doctor on duty objected, knowing that he could breathe only on his left side. Jiang had him move Mao nonetheless. As a result, Mao's breathing stopped and his face turned blue. Jiang Qing left the room while the medical staff put Mao on a respirator and performed emergency cardiopulmonary resuscitation. Eventually, Mao was revived and Hua Guofeng urged Jiang Qing not to interfere further with the doctor's work. However, Mao's organs failed and the Chinese government decided to disconnect Mao's life support mechanism.

Mao's death on September 9, 1976, sent shockwaves through the country. As the symbol of China's revolution, Mao was held in high regard amongst the majority of the Chinese population. Mao's chosen successor, Hua Guofeng, chaired his funeral committee. It was believed Hua was a compromise candidate between the free-marketeers and the party orthodox. Some argue this may have been due to his ambivalence and his low-key profile, particularly compared to Deng Xiaoping, the preferred candidate of the market-oriented factions. The party apparatus, under orders from Jiang Qing and Zhang Chunqiao, wrote a eulogy affirming Mao's achievements and in order to justify their claims to power. By this time state media was effectively under the control of the Gang of Four. State newspapers continued to denounce Deng shortly after Mao's death. Jiang Qing was especially paranoid of Deng's influence on national affairs, whereas she considered Hua Guofeng a mere nuisance. In numerous documents published in the 1980s it was claimed that Jiang Qing was conspiring to make herself the new Chairman of the Communist Party.

Death

Jiang Qing was sentenced to death in 1981. In 1983, her death sentence was commuted to life imprisonment. While in prison, Jiang Qing was diagnosed with throat cancer, but she refused an operation. She was eventually released, on medical grounds, in 1991. At the hospital, Jiang Qing used the name LĐ Rùnqîng. She was alleged to have committed suicide on May 14, 1991, aged 77, by hanging herself in a bathroom of her hospital.

53

Hildegard of Bingen

Blessed Hildegard of Bingen (1098 – 17 September 1179), also known as Saint Hildegard, and Sibyl of the Rhine, was a writer, composer, philosopher, Christian mystic, German Benedictine abbess, visionary, and polymath. Elected a *magistra* by her fellow nuns in 1136, she founded the monasteries of Rupertsberg in 1150 and Eibingen in 1165. One of her works as a composer, the *Ordo Virtutum*, is an early example of liturgical drama. She wrote theological, botanical and medicinal texts, as well as letters, liturgical songs, poems, and arguably the oldest surviving morality play, while supervising brilliant miniature Illuminations.

Biography

Hildegard of Bingen's date of birth is uncertain. It has been concluded that she may have been born in the year 1098. Hildegard was raised in a family of free nobles. She was her parents' tenth child, sickly from birth. In her *Vita*, Hildegard explains that from a very young age she had experienced visions. Perhaps due to Hildegard's visions, or as a method of political positioning, Hildegard's parents, Hildebert and Mechthilde, offered her as a tithe to the church. The date of Hildegard's enclosure in the church is contentious. Her *Vita* tells us she was enclosed with an older nun, Jutta, at the age of eight. However, Jutta's enclosure date is known to be in 1112, at which time Hildegard would have been fourteen. Some scholars speculate that Hildegard was placed in the care of Jutta, the daughter of Count Stephan II of Sponheim, at the age of eight, before the two women were enclosed together six years later. There is no written record of the twenty-four years of Hildegard's life that she was in the convent together with Jutta. It is possible that Hildegard could have been a chantress and a worker in the herbarium and infirmarium. In any case, Hildegard and Jutta were enclosed at Disibodenberg in the Palatinate Forest in what is now Germany. Jutta was also a visionary and thus attracted many followers who came to visit her at the enclosure. Hildegard also tells us that Jutta taught her to read and write, but that she was unlearned and therefore incapable of teaching Hildegard Biblical interpretation. Hildegard and Jutta most likely prayed, meditated, read scriptures such as the psalter, and did some sort of handwork during the hours of the Divine Office. This also might have been a time when Hildegard learned how to play the ten-stringed psaltery. Volmar, a frequent visitor, may have taught Hildegard simple psalm notation. The time she studied music could also have been the beginning of the compositions she would later create.

Upon Jutta's death in 1136, Hildegard was unanimously elected as "magistra" of her sister community by her fellow nuns. Abbot Kuno, the Abbot of Disibodenberg, also asked Hildegard to be Prioress. Hildegard, however, wanted more independence for herself and her nuns and asked Abbot Kuno to allow them to move to Rupertsberg. When the abbot declined Hildegard's proposition, Hildegard went over his head and received the approval of Archbishop Henry I of Mainz. Abbot Kuno did not relent, however, until Hildegard was stricken by an illness that kept her paralyzed and unable to move from her bed, an event that she attributed to God's unhappiness at her not following his orders to move her nuns to Rupertsberg. It was only when the Abbot himself could not move Hildegard that he decided to grant the nuns their own monastery. Hildegard and about twenty nuns thus moved to the St. Rupertsberg monastery in 1150, where Volmar served as provost, as well as Hildegard's confessor and scribe. In 1165 Hildegard founded a second convent for her nuns at Eibingen.

Works

Attention in recent decades to women of the medieval Church has led to a great deal of popular interest in Hildegard, particularly her music. Between 70 and 80 compositions have survived, which is one of the largest repertoires among medieval composers. Hildegard left behind over 100 letters, 72 songs, 70 poems, and 9 books. One of her better known works, *Ordo Virtutum* (*Play of the Virtues*), is a morality play. It is unsure when some of Hildegard's compositions were composed, though the Ordo Virtutum is thought to have been composed as early as 1151. The morality play consists of monophonic melodies for the Anima (human soul) and 16 Virtues. There is also one speaking part for the Devil. Scholars assert that the role of the Devil would have been played by Volmar, while Hildegard's nuns would have played the parts of Anima and the Virtues.

Significance

Hildegard communicated with popes such as Eugene III and Anastasius IV, statesmen such as Abbot Suger, German emperors such as Frederick I Barbarossa, and other notable figures such as Saint Bernard of Clairvaux, who advanced her work, at the behest of her abbot, Kuno, at the Synod of Trier in 1147 and 1148. Hildegard of Bingen's correspondence with many people is an important element of her literary work because this is where we can see her speaking most directly to us. Many abbots and abbesses asked her for prayers and opinions on various matters. She traveled widely during her four preaching tours. She had several rather fanatic followers, including Guibert of Gembloux, who wrote frequently to Hildegard and eventually became her secretary after Volmar died in 1173. In addition, Hildegard influenced several monastic women of her time and the centuries that followed; in particular, she engaged in correspondence with another nearby visionary, Elisabeth of Schonau.

Contributing to Christian European rhetorical traditions, she "authorized herself as a theologian" through alternative rhetorical arts. Hildegard was creative in her interpretation of theology. She believed that her monastery should not allow novices who were from a different class than nobility because it put them in an inferior position. She also stated that 'woman may be made from man, but no man can be made without a woman'. Due to church limitation on public, discursive rhetoric, the medieval rhetorical arts included: preaching, letter writing, poetry, and the encyclopedic tradition. Hildegard's participation in these arts speaks to her significance as a female rhetorician, transcending bans on women's social participation and interpretation of scriptures. The acceptance of public preaching by a woman, even a well-connected abbess and acknowledged prophet does not fit the usual stereotype of this time. She conducted four preaching tours throughout Germany, speaking to both clergy and laity in chapter houses and in public, mainly denouncing clerical corruption and calling for reform. Maddocks claims that it is likely she learned simple Latin, and the tenets of the Christian faith, but was not instructed in the Seven Liberal Arts, which formed the basis of all education for the learned classes in the Middle Ages: the *Trivium* of grammar, dialectic, and rhetoric plus the *Quadrivium* of arithmetic, geometry, astronomy, and music. The correspondence she kept with the outside world both spiritual and social transgressed the cloister as a space of female confinement, and served to document Hildegard's grand style and strict formatting of medieval letter writing.

Recent scholars have asserted that Hildegard made a close association between music and the female body in her musical compositions. The poetry and music of Hildegard's Symphonia is concerned with the anatomy of female desire thus described as Sapphonic, or pertaining to Sappho, connecting her to a history of female rhetoricians. Hildegard was one of the first persons for whom the canonization process was officially applied, but the process took so long that four attempts at canonization were not completed, and she remained at the level of her beatification. Hildegard's name was nonetheless taken up in the Roman Martyrology at the end of the sixteenth century. Her feast day is September 17. Numerous popes have referred to Hildegard as a saint, including Pope John Paul II and Pope Benedict XVI. Hildegard's Parish and Pilgrimage Church house the relics of Hildegard, including an altar encasing her remains, in Eibingen near Rüdesheim. Hildegard of Bingen also appears in the calendar of saints in various Anglican churches. In the Church of England she is commemorated on 17 September.

54

Billie Holiday

Billie Holiday (April 7, 1915 – July 17, 1959) was an American jazz singer and songwriter. Nicknamed Lady Day by her loyal friend and musical partner, Lester Young, Holiday was a seminal influence on jazz and pop singing. Her vocal style, strongly inspired by jazz instrumentalists, pioneered a new way of manipulating phrasing and tempo. Above all, she was admired for her deeply personal and intimate approach to singing. She co-wrote a few songs, and several of them have become jazz standards, notably "God Bless the Child", "Don't Explain", and "Lady Sings the Blues". She also became famous for singing jazz standards written by others, including "Easy Living" and "Strange Fruit." Her early career is hard to track down with accuracy. But, it is said that from an early age she worked as a prostitute for a Harlem brothel and was imprisoned for a short period of time for soliciting.

> I never had a chance to play with dolls like other kids. I started working when I was six years old.
>
> —*Bilie Holiday*

She later gained work singing in local jazz clubs before being spotted by a talent scout, John Hammond, in 1933, aged 18. Her voice and recordings are loved for the depth of emotion and intensity she could bring to classic standards. Her range of voice was not the greatest, but, her distinctive gravelly voice was soon to become very famous and influential. She was an important icon of the jazz era and influential in the development of jazz singing. In the late 1930s she began singing a civil rights song called "Strange Fruit" - a song which told the tale of a lynching of a black man in the deep south.

It was very controversial for that period and it was not played on radios. It was recorded for Commodore records and she performed it many times over the next 20 years. Billie Holiday had a difficult upbringing which influenced her perspective on life. She was drawn to abusive men and she experienced many tempestuous relationships. She also became increasingly dependent on various narcotics which contributed to her premature death in 1959, aged just 44. Citation : Pettinger, Tejvan. "Biography of Bilie Holiday", Oxford, www.biographyonline.net, 28th May. 2010

55

Eva Peron

> "I have one thing that counts, and that is my heart; it burns in my soul, it aches in my flesh, and it ignites my nerves: that is my love for the people and Peron."
>
> —*Eva Peron*

Eva Peron is Argentina's most famous women. She inspired millions with her campaigns to help the poor and give women the right to vote. To her supporters she was a saint who strove to overcome poverty and injustice. To her detractors she was a controversial figure at the heart of Argentinean politics. Eva Peron was born in rural poverty in a town called Los Toldos. She was the illegitimate daughter of a failed land owner. Aged fifteen, she left her rural home to go to Buenos Aires where she hoped to pursue her theatrical career. Aided by her natural beauty she gained work in the theatre. In Buenos Aires she also began campaigning for women to be given the vote and to deal with the widespread poverty endemic in Argentina. She caught the eye of prominent politician, Juan Domingo Peron, and in 1945 they were married and six months later she became President Peron's first lady. As the president's wife she took a high profile in campaigning for issues such as women's rights and for the improvement of the descamisados (shirtless) i.e. the very poor.

> "The nation's government has just handed me the bill that grants us our civil rights. I am receiving it before you, certain that I am accepting this on behalf of all Argentinean women, and I can feel my hands tremble with joy as they grasp the laurel proclaiming victory."
>
> —*Eva Peron*

Her high profile, beauty and concern for the poor galvanised the nation. The dispossessed saw her as a saviour. The military and upper echelons of society saw her as a threat. They criticised her professed concern for the poor as a way to gain support for her President. Others have criticised the regime of Juan Peron for having fascist tendencies - though these remain controversial. Her supporters dispute these assertions of her enemies, arguing the military and political opponents were merely trying to tarnish her image. In 1952 she was given the title of 'spiritual chief of the nation'. Six months later, in 1952, she died tragically young from cancer. In 1955, Juan Peron was overthrown by a military coup, they took her body and had it interned in a Milanese grave under the name of a nun. They feared her legacy would provide a point of opposition to the military regime. In 1973, Juan Peron returned to Argentina to begin a third term as president after the military regime were overthrown. Eva Peron's body was returned in November 1974. Her life was made into a hit musical - *Evita* by Andrew Lloyd Weber and Tim Rice. In 1995, Madonna starred as Evita in a high profile screen version. Eva Peron remains an important symbol of emancipation, especially for women in Latin America. She was one of the first women to create a lasting political / humanitarian legacy. Christina Fernandez, the first female elected President of Argentina, claims that women of her generation owe a debt to Eva for "her example of passion and combativeness".

56

Yoko Ono

Yoko Ono (born February 18, 1933) is a Japanese artist, musician, author and peace activist, known for her work in avant-garde art, music and filmmaking as well as her marriage to John Lennon. Ono brought feminism to the forefront through her music which prefigured New Wave music (whether she was a direct influence is still debated). She is a supporter of gay rights and is known for her philanthropic contributions to the arts, peace and AIDS outreach programs.

Early Life

Yoko Ono was born in 1933 to mother Isoko Ono, the great-granddaughter of Zenjiro Yasuda of the Yasuda banking family, and to father Yeisuke Ono, a banker and one-time classical pianist who was a descendant of an Emperor of Japan. Two weeks before she was born, her father was transferred to San Francisco by his employer, the Yokohama Specie Bank. The rest of the family followed soon after and Yoko met her father when she was two. Her younger brother Keisuke was born in December 1936. In 1937, her father was transferred back to Japan and Ono was enrolled at Tokyo's Gakushuin (also known as the Peers School), one of the most exclusive schools in Japan. In 1940, the family moved to New York City, where Ono's father was working. In 1941, her father was transferred to Hanoi and the family returned to Japan. Ono was then enrolled in Keimei Gakuen, an exclusive Christian primary school run by the Mitsui family. She remained in Tokyo through the great fire-bombing of March 9, 1945. During the fire-bombing, she was sheltered with other members of her family in a special bunker in the Azabu district of

Tokyo, far from the heavy bombing. After the bombing, Ono went to the Karuizawa mountain resort with members of her family.

Ono has said that she and her family were forced to beg for food while pulling their belongings in a wheelbarrow; and it was during this period in her life that Ono says she developed her "aggressive" attitude and understanding of "outsider" status when children taunted her and her brother, who were once well-to-do. Other stories have her mother bringing a large number of goods with them to the countryside which they bartered for food. One famous anecdote has her mother bartering a German-made sewing machine for sixty kilograms of rice with which to feed the family. Her father remained in the city and, unbeknownst to them, was believed to have been eventually incarcerated in a prisoner of war camp in China. In an interview by *Democracy Now*'s Amy Goodman on October 16, 2007, Ono explained, "He was in French Indochina which is Vietnam actually... in Saigon. He was in a concentration camp." By April 1946, Gakushuin was reopened and Ono was enrolled. The school located near the imperial palace, had not been damaged by the war. She graduated in 1951 and was accepted into the philosophy program of Gakushuin University, the first woman to enter the department. However, after two semesters, she left the school.

Education and Marriages

Ono's family moved to Scarsdale, New York after the war. She left Japan to rejoin the family and enrolled in nearby Sarah Lawrence College. While her parents approved of her college choice, they were dismayed at her lifestyle, and, according to Ono, chastised her for befriending people they considered to be "beneath" her. In spite of this, Ono loved meeting artists, poets and others who represented the "bohemian" freedom she longed for herself. Visiting galleries and art "happenings" in the city whetted her desire to publicly display her own artistic endeavors. La Monte Young, her first important contact in the New York art world, helped Ono start her career by using her Chambers Street loft in Tribeca as a performance space. At one performance, Ono set a painting on fire; fortunately John Cage had advised her to treat the paper with flame retardant. In 1956, she married composer Toshi Ichiyanagi. They divorced in 1962 after living apart for several years. On November 28 that same year, Ono married an American named Anthony Cox. Cox was a jazz musician, film producer and art promoter. He had heard of Ono in New York and tracked her down to a mental institution in Japan, where her family had placed her following a suicide attempt. Ono had neglected to finalize her divorce from Ichiyanagi, so their marriage was annulled on March 1, 1963 and Cox and Ono married on June 6. Their daughter, Kyoko Chan Cox, was born two months later on August 8, 1963.

The marriage quickly fell apart but the Coxes stayed together for the sake of their joint career. They performed at Tokyo's Sogetsu Hall with Ono lying atop a piano played by John Cage. Soon the Coxes returned to New York with Kyoko. In the early years of this marriage, Ono left most of Kyoko's parenting to Cox while she pursued her art full-time and Tony managed publicity. After she divorced Cox on February 2, 1969, Ono and Cox engaged in a bitter legal battle for custody of Kyoko, which resulted in Ono being awarded full custody. However, in 1971, Cox disappeared with eight-year-old Kyoko, in violation of the custody order.

Artwork

Ono was a sometime member of Fluxus, a loose association of Dada-inspired avant-garde artists that developed in the early 1960s. Fluxus founder George Maciunas, a friend of Ono's during the 1960s, admired her work and promoted it with enthusiasm. One of Ono's well known examples is when she took a fly as her alter ego and was inspired by this for her work. Maciunas invited Ono to join the Fluxus group, but she declined because she wanted to remain an independent artist. John Cage was one of the most important influences on Ono's performance art. It was her relationship to Ichiyanagi Toshi, who was a pupil of John Cage's legendary class of Experimental Composition at the New School, that would introduce her to the unconventional avant-garde, neo-Dadaism of John Cage and his protégés in New York City.

Ono was also an experimental filmmaker who made sixteen films between 1964 and 1972, and gained particular renown for a 1966 Fluxus film called simply *No. 4*, but often referred to as "Bottoms". The film consists of a series of close-ups of human buttocks as the subject walks on a treadmill. The screen is divided into four almost equal sections by the elements of the gluteal cleft and the horizontal gluteal crease. The soundtrack consists of interviews with those who are being filmed as well as those considering joining the project.

In 2001, *YES YOKO ONO*, a forty-year retrospective of Ono's work, received the prestigious International Association of Art Critics USA Award for Best Museum Show Originating in New York City, considered one of the highest accolades in the museum profession. In 2002 Ono was awarded the Skowhegan Medal for work in assorted media. In 2005 she received a lifetime achievement award from the Japan Society of New York. Ono received an honorary Doctorate of Laws from Liverpool University in 2001. In 2002, she was presented with the honorary degree of Doctor of Fine Arts from Bard College. In 2008, she showed a large retrospective exhibition, Between The Sky And My Head, at the Kunsthalle Bielefeld, Bielefeld, Germany, and the Baltic Centre for Contemporary Art in Gateshead, UK.

Memorials

Ono funded the construction and maintenance of the Strawberry Fields memorial in New York City's Central Park, across from where they lived and Lennon died. It was officially dedicated on October 9, 1985, what would have been his 45th birthday. In 2000, she founded the John Lennon Museum in Saitama, Saitama, Japan. On October 9, 2007, Ono dedicated a new memorial called the Imagine Peace Tower, located on the island of Viðey, 1 km outside the Skarfabakki harbour in Reykjavík in Iceland. Each year, between October 9 and December 8, it projects a vertical beam of light high into the sky. In 2009, Ono created an exhibit called John Lennon: THE NEW YORK CITY YEARS for the NYC Rock and Roll Hall of Fame Annex. The exhibit uses music, photographs and personal items to depict Lennon's life in New York. A portion of the cost of each ticket to the exhibition is donated to Spirit Foundation, a charitable foundation set up by Lennon and Ono.

Musical Career

Ono collaborated with experimental luminaries such as John Cage and jazz legend Ornette Coleman. In 1961, years before meeting Lennon, she had her first major public performance in a concert at the 258-seat Carnegie Recital Hall (not the larger "Main Hall"). This concert featured radical experimental music and performances. She had a second engagement at the Carnegie Recital Hall in 1965, in which she debuted "Cut Piece." In early 1980, Lennon heard Lene Lovich and The B-52's' "Rock Lobster" in a nightclub, and it reminded him of Ono's musical sound. He took this as an indication that her sound had reached the mainstream. Indeed, many musicians, particularly those of the new wave movement, have paid tribute to Ono (both as an artist in her own right, and as a muse and iconic figure). For example, Elvis Costello recorded a version of Ono's song "Walking on Thin Ice," the B-52's who drew from her early recordings covered "Don't Worry, Kyoko (Mummy's Only Looking for Her Hand in the Snow)" (shortening the title to "Don't Worry") and Sonic Youth included a performance of Ono's early conceptual "Voice Piece for Soprano" in their experimental album *SYR4: Goodbye 20th Century*.

Discography

Albums

- *Unfinished Music No.1: Two Virgins* [*] (1968) #124
- *Unfinished Music No.2: Life with the Lions* [*] (1969) #174
- *Wedding Album* [*] (1969) #178
- *Yoko Ono/Plastic Ono Band* (1970) #182
- *Fly* (1971) #199
- *Some Time in New York City* [*] (1972) #48
- *Approximately Infinite Universe* (1972) #193
- *Feeling the Space* (1973)

57

Betty Friedan

Betty Friedan (February 4, 1921 - February 4, 2006) was an American writer, activist, and feminist. A leading figure in the Women's Movement in the United States, her 1963 book *The Feminine Mystique* is often credited with sparking the "second wave" of American feminism in the twentieth century. In 1966, Friedan founded and was elected the first president of the National Organization for Women, which aimed to bring women "into the mainstream of American society now [in] fully equal partnership with men". In 1970, after stepping down as NOW's first president, Friedan organized the nation-wide Women's Strike for Equality on August 26, the 50th anniversary of the Nineteenth Amendment to the United States Constitution granting women the right to vote. The national strike was successful beyond expectations in broadening the feminist movement; the march led by Friedan in New York City alone attracted over 50,000 women and men. In 1971, Friedan joined other leading feminists to establish the National Women's Political Caucus. Friedan was also a strong supporter of the proposed Equal Rights Amendment to the United States Constitution that passed the United States House of Representatives (by a vote of 354-24) and Senate (84-8) following intense pressure by women's groups led by NOW in the early 1970s.

Following Congressional passage of the amendment Friedan advocated for ratification of the amendment in the states and supported other women's rights reforms. Friedan was a strong proponent of the repeal of abortion laws, founding the National Association for the Repeal of Abortion Laws but was later critical of the abortion-centered, politicized tactics of many liberal and radical feminists. Regarded as an influential author and intellectual in the United States, Friedan remained active in politics and advocacy for the rest of her life, authoring six books. As early as the 1960s Friedan was critical of polarized and extreme factions of feminism that attacked groups such as men and homemakers. One of her later books, *The Second Stage*, critiqued what Friedan saw as the extremist excesses of some feminists who could be broadly classified as gender feminists.

Early Life

Freidan was born Bettye Naomi Goldstein on February 4, 1921 in Peoria, Illinois, to Harry and Miriam Goldstein. Harry owned a jewelry store in Peoria, and Miriam wrote for the society page of a newspaper when Friedan's father fell ill. Her mother's new life outside the home seemed much more gratifying. As a young girl, Friedan was active in Marxist and Jewish circles; she later wrote how she felt isolated from the community at times, and felt her "passion against injustice...originated from my feelings of the injustice of anti-Semitism". She attended Peoria High School where she became involved in the school newspaper. When she was turned down for a column, she and six other friends launched a literary magazine called *Tide*. In this magazine, Friedan and her friends talked about home life as opposed to school life. She attended the all-female Smith College in 1938. She won a scholarship prize in her first year for outstanding academic performance. In her second year, she became interested in poetry, and had many poems published in campus publications. In 1941, she became editor-in-chief of the college newspaper. The editorials became more political under her leadership, taking a strong anti-war stance and occasionally causing controversy. She graduated *summa cum laude* in 1942, majoring in psychology.

In 1943, she spent a year at the University of California, Berkeley having won a fellowship to undertake graduate work in psychology with Erik Erikson. She became more politically active, continuing to mix with Marxists (many of her friends were investigated by the FBI). Friedan claims in her memoirs that her boyfriend at the time pressured her into turning down a Ph.D fellowship for further study and abandoning her academic career.

Writing Career

After leaving Berkeley, Friedan became a journalist for leftist and union publications. Between 1943-46 she wrote for *The Federated Press* and between 1946-52 she worked for the United Electrical Workers' *UE News*. One of her assignments was to report on the House Un-American Activities Committee. Friedan was dismissed from the union newspaper *UE News* in 1952, because she was pregnant with her second child. After leaving UE News, she became a freelance writer, and wrote for various magazines, including Cosmopolitan.

For her 15th college reunion in 1957, Friedan conducted a survey of College graduates, focusing on their education, their subsequent experiences and satisfaction with their current lives. She started publishing articles about what she called "the problem that has no name," and got passionate responses from many housewives grateful that they were not alone in experiencing this problem. Friedan then decided to rework and expand this topic into a book, *The Feminine Mystique*. Published in 1963, it depicted the roles of women in industrial societies, especially the full-time homemaker role, which Friedan deemed stifling. Friedan speaks of her own 'terror' at being alone, and observes in her life never once seeing a positive female role-model who worked *and* also kept a family. She provides numerous accounts of housewives who feel similarly trapped. With her psychology background, Friedan offers a critique of Freud's penis envy theory, noting a lot of paradoxes in his work. And she attempts to offer some answers to women who wish to pursue an education. Friedan noted that women are as capable as men to do any type of work or follow any career path, and the mass media, educators, and psychologists argued to the contrary. The restrictions of the 1950s, and the trapped, imprisoned, feeling of many women forced into these roles, spoke to American women who soon began attending consciousness-raising sessions and lobbying for the reform of oppressive laws and social views that restricted women. The book became a bestseller, which many historians believe was the impetus for the "second wave" of the Women's Movement, and significantly shaped national and world events. Friedan published six books. Her other books include *The Second Stage*, *It Changed My Life: Writings on the Women's Movement*, *Beyond Gender*, and *The Fountain of Age*. Her autobiography, *Life so Far*, was published in 2000.

Activism in the Women's Movement

In 1966 Friedan co-founded, and became the first president of, the National Organization for Women. She, with Pauli Murray, the first black female Episcopal priest, wrote its mission statement. Friedan stepped down as president in 1969. Under Friedan, NOW advocated fiercely for the legal equality of women and men. They lobbied for enforcement of Title VII of the Civil Rights Act of 1964 and the Equal Pay Act of 1963, the first two major legislative victories of the movement, and forced the Equal Employment Opportunity Commission to stop ignoring, and start treating with dignity and urgency, claims filed involving sex discrimination. They successfully campaigned for a 1967 Executive Order extending the same Affirmative Action granted to blacks to women and a 1968 EEOC decision ruling illegal sex-segregated help want ads, later upheld by the Supreme Court. NOW was vocal in support of the legalization of abortion, something that divided some feminists. Also divisive in the 1960s among women was the Equal Rights Amendment, which NOW fully endorsed; by the 1970s the women and labor unions opposed to ERA warmed up to it and began to fully support it. NOW also lobbied for national day-care. In 1973, Friedan founded the First Women's Bank and Trust Company.

In 1970, NOW, with Friedan leading the cause, was instrumental in bringing down the nomination of G. Harrold Carswell, who had opposed the 1964 Civil Rights Act which granted women and men workplace equality, to the Supreme Court. On August 26, 1970, the 50th anniversary of the Women's Suffrage Amendment to the Constitution, Friedan organized the national Women's Strike for Equality, and led a march of 50,000 women in New York City. Unbelievably successful, the march expanded the movement widely, to Friedan's delight. Friedan founded the National Association for the Repeal of Abortion Laws, renamed National Abortion Rights Action

League after the Supreme Court legalized abortion in 1973. In 1971 Friedan, along with countless other leading women's movement leaders, including Gloria Steinem, with whom she had a legendary rivalry, founded the National Women's Political Caucus. In 1970 Friedan led other feminists in derailing the nomination of Supreme Court nominee G. One of the most influential feminists of 20th century, Friedan opposed equating feminism with lesbianism. As early as 1964, very early in the movement, and only a year after the publication of *The Feminine Mystique*, Friedan appeared on television to address the fact the media was, at that point, trying to dismiss the movement as a joke and centering argument and debate around whether or not to wear bras and other issues considered ridiculous. In 1982, during the second wave, she wrote a book for the post-feminist 1980s called *The Second Stage*, about family life, premised on women having conquered social and legal obstacles. She pushed the feminist movement to focus on economic issues, especially equality in employment and business and provision for child care and other means by which women and men could balance family and work. She tried to lessen the focus on abortion, as an issue already won, and rape and pornography, which she believed most women did not consider to be high priorities.

Related Issues

Lesbian Politics

When she grew up in Peoria, Ill., she knew one gay man. She said, "the whole idea of homosexuality made me profoundly uneasy." She later acknowledged that she had been very square and was uncomfortable about homosexuality. "The women's movement was not about sex, but about equal opportunity in jobs and all the rest of it. Yes, I suppose you have to say that freedom of sexual choice is part of that, but it shouldn't be the main issue . . . ." She ignored Lesbians in the National Organization for Women (NOW) initially but objected to what she saw as demands for equal time. "'Homosexuality . . . is not, in my opinion, what the women's movement is all about.'" While opposing all repression, she wrote, she refused to wear a purple armband or self-identify as a Lesbian (although heterosexual) as an act of political solidarity, considering it not part of the mainstream issues of abortion and child care. In 1977, at the National Women's Conference, she seconded a Lesbian rights resolution "which everyone thought I would oppose" in order to "preempt any debate" and move on to other issues she believed were more important and less divisive in the effort to add the Equal Rights Amendment (ERA) to the U.S. Constitution. She accepted Lesbian sexuality ("'Enjoy!'"), albeit not its politicization. In 1995, at the United Nations Fourth World Conference on Women, in Beijing, China, she found Chinese advice to taxi drivers that naked Lesbians would be "cavorting" in their cars and so drivers should hang sheets and that Lesbians would have AIDS and so drivers should have disinfectants to be "ridiculous", "incredibly stupid", and "insulting". In 1997, she wrote that "children . . . will ideally come from mother and father." She wrote in 2000, "I'm more relaxed about the whole issue now[.]"

Abortion Choice

She supported the concept that abortion is a woman's choice, that it shouldn't be a crime or exclusively a doctor's choice, and helped form NARAL (now NARAL Pro-Choice America) at a time when Planned Parenthood wasn't yet in support. Death threats against her speaking on abortion led to two events being canceled, although subsequently one of the host institutions, Loyola College, invited her back to speak on abortion and other issues and she spoke then. Her draft of NOW's first statement of purpose included an abortion plank but NOW didn't include it until the next year. In 1980, she believed abortion should be in the context of "'the choice to have children'", a formulation supported by the Roman Catholic priest organizing Catholic participation in the White House Conference on Families of that year, although perhaps not by the bishops above him. A resolution embodying the formulation passed at the conference by 460 to 114, whereas a resolution addressing abortion, ERA, and "'sexual preference'" passed by only 292–291 and that only after 50 anti-abortion advocates had walked out and so hadn't voted on it. She disagreed with a resolution that framed abortion in more feminist terms that was introduced in the Minneapolis regional conference of the same White House Conference on Families, believing it to be more polarizing, while the drafters apparently thought Friedan's formulation too conservative. As of 2000, she wrote, referring to "NOW and the other women's organizations" as seeming to be in a "time warp", "To my mind, there is far too much focus on abortion. . . . [I]n recent years I've gotten a little uneasy about the movement's narrow focus on abortion as if it were the single, all-important issue for women when it's not" She asked, "Why don't we join forces with all who have true reverence for life, including Catholics who oppose abortion, and fight for the choice to have children?"

Pornography

She joined nearly 200 others in Feminists for Free Expression in opposing the Pornography Victims' Compensation Act. "'[T]o suppress free speech in the name of protecting women is dangerous and wrong,' says Friedan. 'Even some blue-jean ads are insulting and denigrating. I'm not adverse to a boycott but I don't think they should be suppressed.'"

War

In 1968, Friedan signed the "Writers and Editors War Tax Protest" pledge, vowing to refuse tax payments in protest against the Vietnam War.

Influence

Friedan is credited for starting the contemporary feminist movement and writing one of the cornerstones of American feminism. Her activist work and her book *The Feminine Mystique* have been a critical influence to authors, educators, writers, anthropologists, journalists, activists, organizations, unions, and everyday women taking part in the feminist movement. Allan Wolf, in *The Mystique of Betty Friedan* writes: "She helped to change not only the thinking but the lives of many American women, but recent books throw into question the intellectual and personal sources of her work." Although there have been some debates on Friedan's work in *The Feminine Mystique* since its publication, there is no doubt that her work for equality for women was sincere and committed. Judith Hennessee (*Betty Friedan: Her Life*)and Daniel Horowitz, a professor of American Studies at Smith College, have also written about Friedan. Horowitz explored Friedan's engagement with the women's movement before she began to work on her book, *The Feminine Mystique* and argues that Friedan's feminism did not start in the 1950s but rather before that in the 1940s. Focusing his study on Friedan's ideas in feminism rather than on her personal life Horowitz's book connects Friedan to the history of American feminism.

Personal Life

She married Carl Friedan, a theatre-producer, in 1947 while working at UE News. Friedan continued to work after marriage, first as a paid employee and, after 1952, as a freelance journalist. The couple divorced in May 1969. Betty claimed in her memoir, *Life So Far* (2000), that Carl had beaten her during their marriage; friends such as Dolores Alexander recalled having to cover up black eyes from Carl's abuse in time for press conferences (Brownmiller 1999, p. 70). Carl Friedan denied abusing her in an interview with *Time* magazine shortly after the book was published, describing the claim as a "complete fabrication". She later said, on *Good Morning America*, "I almost wish I hadn't even written about it, because it's been sensationalized out of context. My husband was not a wife-beater, and I was no passive victim of a wife-beater. We fought a lot, and he was bigger than me." Carl Friedan died in December 2005. The Friedans had three children: Emily, Daniel and Jonathan. One of their sons, Daniel Friedan, is a noted theoretical physicist. Friedan died of congestive heart failure at her home in Washington, D.C., on February 4, 2006, her 85th birthday.

Bibliography

- *The Feminine Mystique* (1963)
- *It Changed My Life* (1976)
- *The Second Stage* (1981)
- *The Fountain of Age* (1993)
- *Beyond Gender* (1997)
- *Life So Far* (2000)

58

Queen Elizabeth II

Born in 1926 Queen Elizabeth II was just 25 years old when she ascended to the British throne on the death of her father, George IV, in 1952. Throughout her long reign Elizabeth II has sought to discharge her role with an enormous sense of duty. Her reign has experienced profound social and economic change. Within the royal family there have been numerous difficult experiences but despite all difficulties and testing times she has retained the respect of the nation. The 1950s could be seen as the "golden age" for the Queen. It was an era of greater deference to royalty; in particular press intrusion was far less. The young Elizabeth was young and beautiful and the world looked upon her as a fairytale princess. She was crowned Queen Elizabeth II on 2 June 1953. It was the first Royal wedding to be televised and it gripped a nation enjoying the end of post war austerity and also the success of Edmund Hilary reaching peak of Mt Everest.

The 1960s saw remarkable social change. The advent of TV gave the public a greater insight into the life of the Royal family. With increased exposure the distance between the royal family and the public diminished. It created a real appetite for greater knowledge and information about royalty. This frenzy of media interest would later create real problems for the Royal family. In the 1970s the queen celebrated her silver jubilee to widespread public interest and support. In the 1980s the focus of the royal family switched to Charles and Diana. In particular the media became increasingly fascinated with the Fairytale princess – Princess Diana. The 1990s saw a catalogue of problems for the queen and Elizabeth. In one Christmas message she referred memorably to an "annus horribilis" but in reality it would have been appropriate for many years. In particular the acrimonious break up of Princess Diana and Prince Charles divided the Royal family with reports saying the queen was less than sympathetic to princess Diana. The death of Princess Diana was a critical moment in the reign of Queen Elizabeth. The unprecedented outpouring of grief around the world left the queen looking rather aloof. Amidst intense pressure the Queen relented and after 3 days instructed the flag to be raised at half mast over Windsor. In the past 10 years these dark days have been put behind. In 2002 there were very successful Golden Jubilee celebrations with a significant display of affection for both the queen and royal family. The Queen continues with a relentless round of official engagements. Despite nearing an age of 81 she appears remarkably mobile with little intent of passing the throne to Charles early.

59

Margaret Thatcher

Margaret Thatcher was Britain's first female prime minister, who became a pivotal figure in British and world politics. After studying at Sommerville College, Oxford university, Mrs Thatcher progressed through the ranks of the Conservative party to become education minister in Ed Heath's government of the early 1970s. It was as education minister that Mrs Thatcher developed a rather crude nickname of "Maggie Thatcher - the milk snatcher" It was as education secretary that Mrs Thatcher ended free school milk. Even as a minister, Mrs Thatcher proclaimed that Britain would never have a female prime minister. However, contrary to tradition and expectation, Mrs Thatcher was elected Prime Minister in the Conservative landslide of 1979. Mrs Thatcher wasted no time in introducing controversial economic policies. She believed that a harsh implementation of Monetarism was necessary to overcome the economic ills of inflation and low growth, which she blamed on the previous Labour government. However, although she was successful in reducing inflation, deflationary monetary policies caused a serious economic recession, in which unemployment rose to 3 million. Opinion was strongly against many of her policies. In a famous letter to the Times newspaper, 360 economist wrote a letter arguing the government should change its policies immediately. However, in true Thatcher style, she refused. Instead she stood up at the Conservative party conference and stated: "You turn if you want to, but this lady is not for turning." It was characteristic of her whole premiership - fierce in her beliefs and unwavering in her commitment.

In the midst of the recession, the Falklands islands were

invaded by the Argentinean army. Mrs Thatcher sent a British expeditionary force to reclaim the islands. With relatively light casualties (although many hundreds died in the conflict) the islands were retaken. This military victory brought a fillip in support for Thatcher. However, it is worth noting she was criticised for both her decision to sink the Belgrano (which was sailing away from the conflict zone) Others also criticised her triumphalist spirit. On reclaiming the islands, Mrs Thatcher proclaimed "rejoice, rejoice" Many felt this was inappropriate given the recent casualties on both the British and Argentinean sides. Another defining feature of the early Thatcher administration was her battle with trades unions. Thatcher wanted to reduce the power of trades unions, in particular, she wished to reduce the influence of the militant mine workers union, the NUM led by Arthur Scargill. Mrs Thatcher prepared the country for a long strike; when the miners went all out on strike in 1984, they were eventually forced back into work after a year long bitter struggle.

The remaining years of her premiership were overshadowed by her controversial and dogmatic decision to stick with the poll tax. This was widely regarded as an unfair tax because everybody paid the same amount regardless of income. Opposition to the poll tax spilled over into violent protest and her popularity plummeted. Because of her declining popularity she was eventually forced out in 1990. Although she was bitter about her perceived betrayal, she left an unprecedented mark on the UK economic and political landscape. For good or ill, she changed the British economic and political situation. It is ironic that when Labour eventually regained power in 1997, it was largely due to the fact Tony Blair and new Labour took on board many of the economic policies that Mrs Thatcher had initiated. There was often a curious mutual respect between Mrs Thatcher and Tony Blair.

60

Dr. Dian Fossey

Dian Fossey is remembered by her fellow scientists as the world's foremost authority on mountain gorillas. But to the millions of wildlife conservationists who came to know Fossey through her articles and book, she will always be remembered as a martyr. Throughout the nearly 20 years she spent studying mountain gorillas in central Africa, the American primatologist tenaciously fought the poachers and bounty hunters who threatened to wipe out the endangered primates. She was brutally murdered at her research center in 1985 by what many believe was a vengeful poacher. Fossey's dream of living in the wilds of Africa dates back to her lonely childhood in San Francisco. She was born in 1932, the only child of George, an insurance agent, and Kitty, a fashion model, (Kidd) Fossey. The Fosseys divorced when Dian was six years old. A year later, Kitty married a wealthy building contractor named Richard Price. Price was a strict disciplinarian who showed little affection for his stepdaughter. Although Fossey loved animals, she was allowed to have only a goldfish. When it died, she cried for a week. Fossey began her college education at the University of California at Davis in the preveterinary medicine program.

61

Germaine Greer

Germaine Greer (born 29 January 1939) is an Australian writer, academic, journalist and scholar of early modern English literature, widely regarded as one of the most significant feminist voices of the later 20th century. Greer's ideas have created controversy ever since her book *The Female Eunuch* became an international best-seller in 1970, turning her into a household name and bringing her both adulation and opposition. She is also the author of many other books including, *Sex and Destiny: The Politics of Human Fertility* (1984); *The Change: Women, Ageing and the Menopause* (1991) and *Shakespeare's Wife* (2007). She is Professor Emeritus of English Literature and Comparative Studies at the University of Warwick. Greer has defined her goal as 'women's liberation' as distinct from 'equality with men', She asserts that women's liberation meant embracing gender differences in a positive fashion, a struggle for the freedom for women to define their own values, order their own priorities and determine their own fates. In contrast, Greer sees equality as mere assimilation and "settling" to live the lives of "unfree men".

Early Life and Career

Germaine Greer was born in Melbourne in 1939, growing up in the bayside suburb of Mentone. Her father was a newspaper advertising rep' who served in the wartime RAAF. After attending a private convent school, Star of the Sea College, in Gardenvale, she won a teaching scholarship in 1956 and enrolled at the University of Melbourne. After graduating with a degree in English and French language and literature, she moved to Sydney, where she became involved with the Sydney Push social milieu and the anarchist Sydney Libertarians at its centre. Christine Wallace, in her unauthorised biography, describes Greer at this time: For Germaine, [the Push] provided a philosophy to underpin the attitude and lifestyle she had already acquired in Melbourne. She walked into the Royal George Hotel, into the throng talking themselves hoarse in a room stinking of stale beer and thick with cigarette smoke, and set out to follow the Push way of life - 'an intolerably difficult discipline which I forced myself to learn'. The Push struck her as completely different from the Melbourne intelligentsia she had engaged with in the Drift, 'who always talked about art and truth and beauty and argument *ad hominem*; instead, these people talked about truth and only truth, insisting that most of what we were exposed to during the day was ideology, which was a synonym for lies - or bullshit, as they called it.' Her Damascus turned out to be the Royal George, and the Hume Highway was the road linking it. 'I was already an anarchist,' she says. 'I just didn't know why I was an anarchist. They put me in touch with the basic texts and I found out what the internal logic was about how I felt and thought.

By 1972 Greer would identify as an anarchist communist, close to Marxism. In her first teaching post, Greer lectured at the University of Sydney, where she also earned a first class MA in romantic poetry in 1963 with a thesis titled *The Development of Byron's Satiric Mode*. A year later, the thesis won her a Commonwealth Scholarship, which she used to fund her doctorate at the University of Cambridge in England, where she became a member of the all-women's Newnham College. Professor Lisa Jardine, who was at Newnham at the same time, recalled the first time she met Greer, at a formal dinner in college: "The principal called us to order for the speeches. As a hush descended, one person continued to speak, too engrossed in her conversation to notice, her strong Australian accent reverberating around the room. At the graduates' table, Germaine was explaining that there could be no liberation for women, no matter how highly educated, as long as we were required to cram our breasts into bras constructed like mini-Vesuviuses, two stitched white cantilevered cones which bore no resemblance to the female anatomy. The willingly suffered discomfort of the Sixties bra, she opined vigorously, was a hideous symbol of male oppression ... [We were] astonished at the very idea that a woman could speak so loudly and out of turn and that words such as "bra" and "breasts' - or maybe she said "tits" - could be uttered amid the pseudo-masculine solemnity of a college dinner.

Following the success of *The Female Eunuch,* Greer resigned her post at Warwick University in 1972 after travelling the world to promote her book. She co-presented a Granada Television comedy show called *Nice Time* with Kenny Everett and Jonathan Routh, bought a house in Italy, wrote a column for *The Sunday Times*, then spent the next few years travelling through Africa and Asia, which included a visit to Bangladesh to investigate the situation of women who had been raped during the conflict with Pakistan. On the New Zealand leg of her tour in 1972, Greer was arrested for using the words "bullshit" and "fuck" during her speech, which attracted major rallies in her support. In the mid-1970s, Greer appeared on conservative William F. Buckley's *Firing Line*. In his memoir, Buckley recalled that Greer had "trounced him" during the debate.

In 1989, Greer was appointed as a special lecturer and fellow at Newnham College, Cambridge. Greer unsuccessfully opposed the election to a fellowship of her transsexual colleague Rachael Padman. Greer argued that Padman had been born male, and therefore should not be admitted to Newnham, a women's college. Greer resigned in 1996 after the case attracted negative publicity. An article concerning the incident was published on 25 June 1997 by Clare Longrigg of *The Guardian*. Entitled "A Sister with No Fellow Feeling"; it disappeared from websites after print publication, on the instruction of the newspaper's lawyers. Greer appeared alongside Daniel O'Donnell the popular Irish crooner on an RTE chat show in 2006. While O'Donnell spoke about love, cooking and his mother's pancakes, Greer mocked him by making faces to camera behind his back. The normally unflappable O'Donnell confronted her and some bitter words were exchanged. Over the years Greer has continued to self-identify as an anarchist or Marxist. In her books she has dealt very little with political labels of this type, but has reaffirmed her position in interviews.

She stated on ABC Television in 2008 that "I ought to confess I suppose that I'm a Marxist. Speaking on an interview for 3CR (an Australian community radio), also in 2008, she described herself as "an old anarchist" and reaffirmed that opposition to "hierarchy and capitalism" were at the centre of her politics.

Works

The Female Eunuch

Greer argued in her book, *The Female Eunuch*, that women do not realise how much men hate them, and how much they are taught to hate themselves. Christine Wallace writes that, when *The Female Eunuch* was first published, one woman had to keep it wrapped in brown paper because her husband wouldn't let her read it; arguments and fights broke out over dinner tables and copies of it were thrown across rooms at unsuspecting husbands (Wallace 1997). It arrived in the shops in London in October 1970. By March 1971, it had nearly sold out its second printing and had been translated into eight languages. "The title is an indication of the problem," Greer told the *New York Times* in 1971, "Women have somehow been separated from their libido, from their faculty of desire, from their sexuality. They've become suspicious about it. Like beasts, for example, who are castrated in farming in order to serve their master's ulterior motives - to be fattened or made docile - women have been cut off from their capacity for action. It's a process that sacrifices vigour for delicacy and succulence, and one that's got to be changed." Two of the book's themes already pointed the way to *Sex and Destiny* 14 years later, namely that the nuclear family is a bad environment for women and for the raising of children; and that the manufacture of women's sexuality by Western society was demeaning and confining. Girls are feminised from childhood by being taught rules that subjugate them, she argued. Later, when women embrace the stereotypical version of adult femininity, they develop a sense of shame about their own bodies, and lose their natural and political autonomy.

Publications in the 1970s and 1980s

Greer's second book, *The Obstacle Race: The Fortunes of Women Painters and Their Work* (1979) covers its subject until the end of the nineteenth century. It also speculates on the existence of women artists whose careers are not recorded by posterity. Greer translated Aristophanes's *Lysistrata* in 1972. *Sex and Destiny: The Politics of Human Fertility*, published in 1984, continued Greer's critique of Western attitudes toward sexuality, fertility, family, and the imposition of those attitudes on the rest of the world. Greer's target again is the nuclear family, government intervention in sexual behaviour, and the commercialisation of sexuality and women's bodies. Germaine Greer argued that the Western promotion of birth control in the Third World was in large part driven not by concern for human welfare but by the traditional fear and envy of the rich towards the fertility of the poor. She argued that the birth control movement had been tainted by such attitudes from its beginning, citing Marie Stopes and others. She cautioned against condemning life styles and family values in the developing world. In 1986, Greer published *Shakespeare*, a work of literary criticism, and *The Madwoman's Underclothes: Essays and Occasional Writings*, a collection of newspaper and magazine articles written between 1968 and 1985. In 1989 came *Daddy, We Hardly Knew You*, a diary and travelogue about her father, whom she described as distant, weak and unaffectionate, which led to claims - which she characterized as inevitable in an interview with *The Guardian* - that in her writing she was projecting her relationship with him onto all other men.

Publications Since 1990

In 1991, *The Change: Women, Ageing, and the Menopause*, which the *New York Times* called a "brilliant, gutsy, exhilarating, exasperating fury of a book" became another influential book in the women's movement. In it, Greer wrote of the various myths concerning menopause, advising against the use of hormone replacement therapy. "Frightening females is fun," she wrote in *The Age*. "Women were frightened intc using hormone replacement therapy by dire predictions of crumbling bones, heart disease, loss of libido, depression, despair, disease and death if they let nature take its course." She argues that scaring women is "big business and hugely profitable." It is fear, she wrote, that "makes women comply with schemes and policies that work against their interest". *Slip-Shod Sibyls:*

Recognition, Rejection and the Woman Poet followed in 1995. In 1999, *the whole woman*, which was intended as a sequel to *The Female Eunuch* was released. In this book she discussed what she saw as the lack of fundamental progress in the feminist movement, and criticized some sections of the women's movement for illusions on that score: "Even if it had been real, equality would have been a poor substitute for liberation; fake equality is leading women into double jeopardy. The rhetoric of equality is being used in the name of political correctness to mask the hammering that women are taking. When *The Female Eunuch* was written our daughters were not cutting or starving themselves. On every side speechless women endure endless hardship, grief and pain, in a world system that creates billions of losers for every handful of winners. It's time to get angry again."Chapter titles reveal the themes, including: "Food," "Breast," "Pantomime Dames (about transsexual women)," "Shopping," "Estrogen," "Testosterone," "Wives," "Loathing," "Girlpower" and "Mutilation" (including a discussion of female genital mutilation in the Third World and the West). Her comments about female genital mutilation proved especially controversial in some quarters, for example a United Kingdom House of Commons Committee described her viewpoint as "simplistic and offensive."

62

Enid Blyton

Enid Blyton (11 August 1897 – 28 November 1968). Enid Blyton (also known as Mary Pollock) was the most successful children's writer of her generation. A prolific writer she completed over 400 books during her lifetime. She is in the top 10 all time best seller lists- her books having sold over 600 million copies Shy in her relations with the public, Enid Blyton revealed little of her private life. She was born in the late 1890s and was brought up in Buckingham. Her father hoped she might become a concert pianist, but despite her talent for music she made the decision to be a children's writer. She took Froebel training and became a governess to a family of boys in Surrey and this experience encouraged her to set up a school for boys. In her spare time she began writing a variety of children's stories. These ranged from natural botany books, biblical stories, a simplified version of Pilgrim's Promise, to the Famous Five series and the ubiquitous Noddy and Big Ears stories. In 1924, she married her first husband - H.A.Pollock with whom she had two daughters. She married her second husband Kenneth Waters in 1943. Her first stories were published by George Newness and her fame grew through the popularity of her stories in the children's magazine 'Sunny Stories.' Her books were controversial amongst literary critics and librarians. Her writings were often not seen as 'great literature' Some found the likes of Big Ears and Noddy just to childish (poor Noddy would frequently burst into tears at any sign of trouble in toytown). For a considerable period, certain libraries would refuse to stock Enid Blyton's books - despite strong demand from children themselves. Yet, whilst the works of Enid Blyton might not have touched the heights of literature - and may compare unfavourably to the more 'adult' success of J.K.Rowling, her books were undoubtedly very popular amongst her core audience and did help a generation of children become interested in reading.

63

Betty Williams

Betty Williams (born 22 May 1943) was a co-recipient with Mairead Corrigan of the Nobel Peace Prize in 1976 for her work as a cofounder of Community of Peace People, an organisation dedicated to promoting a peaceful resolution to The Troubles in Northern Ireland. She heads the Global Children's Foundation and is President of the World Centers of Compassion for Children International. She is also the Chair of Institute for Asian Democracy in Washington D.C. and a Distinguished Visiting Professor at Nova Southeastern University. In 2006, Williams was one of the founders of the Nobel Women's Initiative along with sister Nobel Peace Laureates Mairead Corrigan Maguire, Shirin Ebadi, Wangari Maathai, Jody Williams and Rigoberta Menchu Tum. Six women representing North America and South America, Europe, the Middle East and Africa decided to bring together their experiences in a united effort for peace with justice and equality. It is the goal of the Nobel Women's Initiative to help strengthen work being done in support of women's rights around the world.

- Nobel Peace Prize

First Declaration of the Peace People

- We have a simple message to the world from this movement for Peace.
- We want to live and love and build a just and peaceful society.
- We want for our children, as we want for ourselves, our lives at home, at work, and at play to be lives of joy and Peace.

- We recognize that to build such a society demands dedication, hard work, and courage.
- We recognize that there are many problems in our society which are a source of conflict and violence.
- We recognize that every bullet fired and every exploding bomb make that work more difficult.
- We reject the use of the bomb and the bullet and all the techniques of violence.
- We dedicate ourselves to working with our neighbours, near and far, day in and day out, to build that peaceful society in which the tragedies we have known are a bad memory and a continuing warning

64

Lauren Bacall

Lauren Bacall (born September 16, 1924) is an American film and stage actress and model, known for her distinctive husky voice and sultry looks. She first emerged as leading lady in the film noir genre, including appearances in *The Big Sleep* (1946) and *Dark Passage* (1947), as well as a comedienne in *How to Marry a Millionaire* (1953) and *Designing Woman* (1957). Bacall has also worked in the Broadway musical, gaining Tony Awards for *Applause* in 1970 and *Woman of the Year* in 1981. Her performance in the movie *The Mirror Has Two Faces* (1996) earned her a Golden Globe Award and an Academy Award nomination. In 1999, Bacall was ranked #20 of the 25 actresses on the AFI's 100 Years... 100 Stars list by the American Film Institute. In 2009, she was selected by the Academy of Motion Picture Arts and Sciences to receive an Academy Honorary Award at the inaugural Governors Awards.

Early Life

Born Betty Joan Perske in New York City, Bacall was the only child of Natalie Weinstein-Bacal, a secretary who later legally changed her surname to Bacall, and William Perske, who worked in sales. Her parents were Jewish immigrants, their families having come from Poland, Romania, and Germany. She is first cousin to Shimon Peres, current President and former Prime Minister of Israel. Her parents divorced when she was five, and she took her mother's last name, Bacall. Bacall no longer saw her father and formed a close bond with her mother, whom she took with her to California when she became a movie star.

Career

Bacall took lessons at the American Academy of Dramatic Arts. During this time, she became a theatre usher and worked as a fashion model. As Betty Bacall, she made her acting debut, at age 17, on Broadway in 1942, as a walk-on in *Johnny 2 X 4*. According to her autobiography, she met her idol Bette Davis at Davis' hotel. Years later, Davis visited Bacall backstage to congratulate her on her performance in *Applause*, a musical based on Davis' turn in *All About Eve*. Bacall became a part-time fashion model. Howard Hawks' wife Nancy spotted her on the March 1943 cover of *Harper's Bazaar* and urged Hawks to have her take a screen test for *To Have and Have Not.* Hawks invited her to Hollywood for the audition. He signed her up to a seven-year personal contract, brought her to Hollywood, gave her $100 a week, and began to manage her career. During screen tests for *To Have and Have Not* (1944), Bacall was nervous. To minimize her quivering, she pressed her chin against her chest and to face the camera, tilted her eyes upward. This effect became known as "The Look", Bacall's trademark. On the set, Humphrey Bogart, who was married to Mayo Methot, initiated a relationship with Bacall some weeks into shooting and they began seeing each other.

During the 1980s and early 1990s, Bacall appeared in the poorly received star vehicle *The Fan* (1981), as well as some star-studded features such as Robert Altman's *Health* (1980), Michael Winner's *Appointment with Death* (1988), and Rob Reiner's *Misery* (1990). In 1997, Bacall was nominated for a Best Supporting Actress Academy Award for her role in *The Mirror Has Two Faces* (1996), her first nomination after a career span of more than fifty years. She had already won a Golden Globe and was widely expected to win the Oscar, which went to Juliette Binoche for *The English Patient.* Bacall received the Kennedy Center Honors in 1997. In 1999, she was voted one of the 25 most significant female movie stars in history by the American Film Institute.

Personal Life

On May 21, 1945, Bacall married Humphrey Bogart. Their wedding and honeymoon took place at Malabar Farm, Lucas,

Ohio. It was the country home of Pulitzer Prize-winning author Louis Bromfield, a close friend of Bogart. The wedding was held in the Big House. Bacall was 20 and Bogart was 45. They remained married until Bogart's death from esophageal cancer in 1957. Bogart usually called Bacall "Baby," even when referring to her in conversations with other people. During the filming of *The African Queen* (1951), Bacall and Bogart became friends of Bogart's co-star Katharine Hepburn and her partner Spencer Tracy. Bacall also began to mix in non-acting circles, becoming friends with the historian Arthur Schlesinger, Jr. and the journalist Alistair Cooke. In 1952, she gave campaign speeches for Democratic Presidential contender Adlai Stevenson. Along with other Hollywood figures, Bacall was a staunch opponent of McCarthyism. Shortly after Bogart's death in 1957, Bacall had a relationship with singer and actor Frank Sinatra. She told Robert Osborne, of Turner Classic Movies (TCM), in an interview that she had ended the romance. However, in her autobiography, she wrote that Sinatra abruptly ended the relationship, having become angry that the story of his proposal to Bacall had reached the press. Bacall and her friend Swifty Lazar had run into the gossip columnist Louella Parsons, to whom Lazar had spilled the beans. Sinatra then cut Bacall off and went to Las Vegas. Bacall was married to actor Jason Robards from 1961 to 1969. According to Bacall's autobiography, she divorced Robards mainly because of his alcoholism. In her autobiography *Now*, she recalls having a relationship with Len Cariou, her co-star in *Applause*.

Political Views

Bacall is a staunch liberal Democrat. She has proclaimed her political views on numerous occasions. She appeared alongside Humphrey Bogart in a photograph printed at the end of an article he wrote, titled "I'm No Communist", in the May 1948 edition of *Photoplay* magazine, written to counteract negative publicity resulting from his appearance before the House Un-American Activities Committee. Bogart and Bacall specifically distanced themselves from the Hollywood Ten and were quoted as saying: "We're about as much in favor of Communism as J. Edgar Hoover." In October 1947, Bacall and Bogart traveled to Washington, DC along with other

Hollywood stars, in a group that called itself the Committee for the First Amendment (CFA). She campaigned for Democratic candidate Adlai Stevenson in the 1952 Presidential election and for Robert Kennedy in his 1964 run for Senate. In a 2005 interview with Larry King, Bacall described herself as "anti-Republican.

Dramatization

On Screen and in Literature

- In 1980, Kathryn Harrold played Bacall in the TV movie *Bogie*, which was directed by Vincent Sherman and based on the novel by Joe Hymans. Kevin O'Connor played Bogart. The movie focused primarily upon the disintegration of Bogart's third marriage to Mayo Methot, played by Ann Wedgeworth, when Bogart met Bacall and began an affair with her.

In music

- Bacall is referenced in the song, "Rainbow High", from the musical Evita by Andrew Lloyd Webber and Tim Rice
- Bacall is referenced in the song, "Car Jamming", by 70's punk band The Clash.
- Bacall is also referenced in the song, "Key Largo", by Bertie Higgins.
- She is also referenced in the song "Freeze Tag", by Suzanne Vega.
- She is referenced in the song "Mersey" by Pavlov's Dog (band)
- She is referenced in Madonna's "Vogue".
- She is referenced in the song, "Captain Crash & The Beauty Queen From Mars", by Bon Jovi

In Fashion

- Bacall is paid tribute in "Ariege", one of the latest collections by luxury handbag designer Marcela Calvet

65

Julie Andrews

Dame Julia Elizabeth Andrews, (born 1 October 1935) is a British film and stage actress, singer, and author. She is the recipient of Golden Globe, Emmy, Grammy, BAFTA, People's Choice Award, Theatre World Award, Screen Actors Guild and Academy Award honours. Andrews was a former British child actress and singer who made her Broadway debut in 1954 with *The Boy Friend*, and rose to prominence starring in other musicals such as *My Fair Lady* and *Camelot*, and in musical films such as *Mary Poppins* (1964), for which she won the Academy Award for Best Actress, and *The Sound of Music* (1965): the roles for which she is still best-known. Her voice, which originally spanned four octaves, was damaged by a throat operation in 1997. Andrews had a revival of her film career in 2000s in family films such as *The Princess Diaries* (2001), its sequel *The Princess Diaries 2: Royal Engagement* (2004), the *Shrek* animated films (2004–2010), and *Despicable Me* (2010). In 2003 Andrews revisited her first Broadway success, this time as a stage director, with a revival of *The Boy Friend* at the Bay Street Theatre, Sag Harbor, New York (and later at the Goodspeed Opera House, in East Haddam, Connecticut in 2005). Andrews is also an author of children's books, and in 2008 published an autobiography, *Home: A Memoir of My Early Years*.

Early Life

Julie Andrews was born Julia Elizabeth Wells on 1 October 1935 in Walton-on-Thames, Surrey, England. Her mother, Barbara Ward Wells (née Morris), was married to Edward Charles "Ted" Wells, a teacher of metal and woodworking, but

Andrews was conceived as a result of an affair her mother had with a family friend. With the outbreak of World War II, Barbara and Ted Wells went their separate ways. Ted Wells assisted with evacuating children to Surrey during the Blitz, while Barbara joined Ted Andrews in entertaining the troops through the good offices of the Entertainments National Service Association (ENSA). Barbara and Ted Wells were soon divorced. They both remarried: Barbara to Ted Andrews, in 1939; and Ted Wells, to a former hairstylist working a lathe at a war factory that employed them both in Hinchley Wood, Surrey.

Early Career in the United Kingdom

Julie Andrews performed spontaneously and unbilled on stage with her parents for about two years beginning in 1945. "Then came the day when I was told I must go to bed in the afternoon because I was going to be allowed to sing with Mummy and Pop in the evening," Andrews explained. She would stand on a beer crate to reach the microphone and sing, sometimes a solo or as a duet with her stepfather, while her mother played piano. "It must have been ghastly, but it seemed to go down all right." Julie Andrews got her big break when her stepfather introduced her to Val Parnell, whose Moss Empires controlled prominent venues in London. Andrews made her professional solo debut at the London Hippodrome singing the difficult aria "Je Suis Titania" from *Mignon* as part of a musical revue called "Starlight Roof" on 22 October 1947. She played the Hippodrome for one year. Andrews recalled "Starlight Roof" saying, "There was this wonderful American person and comedian, Wally Boag, who made balloon animals. He would say, 'Is there any little girl or boy in the audience who would like one of these?' And I would rush up onstage and say, 'I'd like one, please.' And then he would chat to me and I'd tell him I sang... I was fortunate in that I absolutely stopped the show cold. I mean, the audience went crazy." On 1 November 1948, Julie Andrews became the youngest solo performer ever to be seen in a Royal Command Variety Performance, at the London Palladium, where she performed along with Danny Kaye, the Nicholas Brothers and the comedy team George and Bert Bernard for members of King George VI's family.

2010–Present

In January 2010, for the second consecutive time, Andrews was the official USA presenter of the New Year's Day Vienna concert. Andrews also had a supporting role in the film *Tooth Fairy*, which opened to unfavourable reviews although the box office receipts were successful. On her promotion tour for the film she also spoke of *Operation USA* and the aid campaign to the Haiti disaster. On 8 May 2010, Andrews made her London comeback after a 21-year absence (her last performance there was a Christmas concert at the Royal Festival Hall in 1989). She performed at the O2 Arena, accompanied by the Royal Philharmonic Orchestra and an ensemble of five performers. Previous to it she appeared on British television (on 15 December 2009 and on many other occasions), and said that rumours that she would be singing were not true. Instead, she said she would be doing a form of "speak singing". However in the concert she actually sang two solos and several duets and ensemble pieces. The evening, though well received by the 20,000 fans present, who gave her standing ovation after standing ovation, did not convince the critics. On 18 May 2010, Andrews' 23rd book (this one also written with her daughter Emma) was published. In June 2010 the book, entitled *The Very Fairy Princess*, reached number 1 on the New York Times Best Seller List for Children's Books.

On 21 May 2010, her film *Shrek Forever After* was released; in it Andrews reprises her role as the Queen. On 9 July 2010, *Despicable Me*, an animated movie in which Andrews lent her voice to Marlena, the evil mother of the main character (Gru, voiced by Steve Carell), opened to rave reviews and strong box office. On 28 October 2010, Andrews appeared, along with the actors who portrayed the cinematic Von Trapp family members, on *Oprah* to commemorate the film's 45th anniversary. A few days later, her 24th book, *Little Bo in Italy*, was published. On 15 December 2010, Andrews' husband Blake Edwards died of complications of pneumonia at the Saint John's Health Center in Santa Monica, California. Andrews was by her husband's side when he passed away.

66

Jane Goodall

Dame Jane Morris Goodall, DBE (3 April 1934), is a British primatologist, ethologist, anthropologist, and UN Messenger of Peace. Considered to be the world's foremost expert on chimpanzees, Goodall is best known for her 45-year study of social and family interactions of wild chimpanzees in Gombe Stream National Park, Tanzania. She is the founder of the Jane Goodall Institute and has worked extensively on conservation and animal welfare issues.

Biography

Jane Goodall was born in London, England in 1934 to Mortimer Herbert Morris-Goodall, a businessman, and Margaret Myfanwe Joseph, a novelist who wrote under the name Vanne Morris-Goodall. As a child she was given a lifelike chimpanzee toy named Jubilee by her father; her fondness for the toy started her early love of animals. Today, the toy still sits on her dresser in London. As she writes in her book, *Reason For Hope*: "My mother's friends were horrified by this toy, thinking it would frighten me and give me nightmares." Goodall had always been passionate about animals and Africa, which brought her to the farm of a friend in the Kenya highlands in 1957. From there, she obtained work as a secretary, and acting on her friend's advice she telephoned Louis Leakey, a Kenyan archaeologist and paleontologist, with no other thought than to make an appointment to discuss animals. Leakey, believing that the study of existing great apes could provide indications of the behaviour of early hominids, was looking for a chimpanzee researcher though he kept the idea to himself. Instead, he proposed that Goodall work for him as a secretary. After obtaining his wife Mary Leakey's approval, Louis sent

Goodall to Olduvai Gorge in Tanzania, where he laid out his plans.

In 1958, Leakey sent Goodall to London to study primate behavior with Osman Hill and primate anatomy with John Napier. Leakey raised funds, and in 1960 Goodall went to Gombe Stream National Park becoming the first of "Leakey's Angels". She was accompanied by her mother whose presence was necessary to satisfy the requirements of David Anstey, chief warden, who was concerned for their safety; Tanzania was *"Tanganyika"* at that time and a British protectorate. Leakey arranged funding and in 1962 sent Goodall, who had no degree, to Cambridge University where she obtained a Ph.D degree in Ethology. She became only the eighth person to be allowed to study for a Ph.D without first obtaining a BA or B.Sc. Her thesis was completed in 1965 under the tutorship of Robert Hinde, former master of St. John's College, Cambridge, titled "Behavior of the Free-Ranging Chimpanzee," detailing her first five years of study at the Gombe Reserve. Goodall has been married twice. On 28 March 1964, she married a Dutch nobleman, wildlife photographer Baron Hugo van Lawick, at Chelsea Old Church, London, and became known during their marriage as Baroness Jane van Lawick-Goodall. The couple had a son, Hugo Eric Louis, affectionately known as "Grub," who was born in 1967. They divorced in 1974. In 1975, she married Derek Bryceson (a member of Tanzania's parliament and the director of that country's national parks); he died of cancer in October 1980. With his position in the Tanzanian government as head of the country's national park system, Bryceson was able to protect Goodall's research project and implement an embargo on tourism at Gombe while he was alive. When asked if she believed in God, Goodall said in September 2010: "I don't have any idea of who or what God is. But I do believe in some great spiritual power. I don't know what to call it. I feel it particularly when I'm out in nature. It's just something that's bigger and stronger than what I am or what anybody is. I feel it. And it's enough for me."

Work

Research at Gombe Stream National Park

Goodall is best known for her study of chimpanzee social

and family life. She began studying the Kasakela chimpanzee community in Gombe Stream National Park, Tanzania in 1960. Without collegiate training directing her research, Goodall observed things that strict scientific doctrines may have overlooked. Instead of numbering the chimpanzees she observed, she gave them names such as Fifi and David Greybeard, and observed them to have unique and individual personalities, an unconventional idea at the time. She found that, "it isn't only human beings who have personality, who are capable of rational thought [and] emotions like joy and sorrow." She also observed behaviors such as hugs, kisses, pats on the back, and even tickling, what we consider "human" actions. Goodall insists that these gestures are evidence of "the close, supportive, affectionate bonds that develop between family members and other individuals within a community, which can persist throughout a life span of more than 50 years." These findings suggest similarities between humans and chimpanzees exist in more than genes alone, but can be seen in emotion, intelligence, and family and social relationships. Goodall's research at Gombe Stream is best known to the scientific community for challenging two long-standing beliefs of the day: that only humans could construct and use tools, and that chimpanzees were vegetarians. While observing one chimpanzee feeding at a termite mound, she watched him repeatedly place stalks of grass into termite holes, then remove them from the hole covered with clinging termites, effectively "fishing" for termites. The chimps would also take twigs from trees and strip off the leaves to make the twig more effective, a form of object modification which is the rudimentary beginnings of toolmaking. Humans had long distinguished ourselves from the rest of the animal kingdom as "Man the Toolmaker". In response to Goodall's revolutionary findings, Louis Leakey wrote, "We must now redefine man, redefine tool, or accept chimpanzees as human!".

Controversy

Some primatologists have suggested flaws in Goodall's methodology which may call into question the validity of her observations. Goodall used unconventional practices in her study, for example, naming individuals instead of numbering them. At the time numbering was used to prevent emotional attachment and loss of objectivity. Claiming to see individuality and emotion in chimpanzees, she was accused of "that worst of ethological sins", anthropomorphism. Many standard methods are aimed at helping observers to avoid interference and the use of feeding stations to attract Gombe chimpanzees is, in particular, thought by some to have altered normal foraging and feeding patterns as well as social relationships; this argument is the focus of a book published by Margaret Power in 1991. It has been suggested that higher levels of aggression and conflict with other chimpanzee groups in the area were consequences of the feeding, which could have created the "wars" between chimpanzee social groups described by Goodall, aspects of which she did not witness in the years before artificial feeding began at Gombe.

Awards and Recognition

Honours

Goodall has received many honors for her environmental and humanitarian work, as well as others. She was named a Dame Commander of the Order of the British Empire in a ceremony held in Buckingham Palace in 2004. In April 2002, Secretary-General Kofi Annan named Goodall a United Nations Messenger of Peace. Her other honors include the Tyler Prize for Environmental Achievement, the French Legion of Honor, Medal of Tanzania, Japan's prestigious Kyoto Prize, the Benjamin Franklin Medal in Life Science, the Gandhi-King Award for Nonviolence and the Spanish Prince of Asturias Awards. She is also a member of the advisory board of *BBC Wildlife* magazine and a patron of Population Matters (formerly the Optimum Population Trust). She has received many tributes, honors, and awards from local governments, schools, institutions, and charities around the world.

Awards

- 1980: Order of the Golden Ark, World Wildlife Award for Conservation
- 1984: J. Paul Getty Wildlife Conservation Prize
- 1985: Living Legacy Award from the International Women's League
- Society of the United States; Award for Humane

Excellence, American Society for the Prevention of Cruelty to Animals

- 1987: Ian Biggs' Prize
- 1989: Encyclopædia Britannica Award for Excellence on the Dissemination of Learning for the Benefit of Mankind; Anthropologist of the Year Award
- 1990: The AMES Award, American Anthropologist Association; Whooping Crane Conservation Award, Conoco, Inc.; Gold Medal of the Society of Women Geographers; Inamori Foundation Award; Washoe Award; The Kyoto Prize in Basic Science
- 1991: The Edinburgh Medal
- 1993: Rainforest Alliance Champion Award
- 1994: Chester Zoo Diamond Jubilee Medal
- 1995: Commander of the Order of the British Empire, presented by Her Majesty Queen Elizabeth II; The National Geographic Society Hubbard Medal for Distinction in Exploration, Discovery, and Research; Lifetime Achievement Award, In Defense of Animals; The Moody Gardens Environmental Award; Honorary Wardenship of Uganda National Parks
- 1996: The Zoological Society of London Silver Medal; The Tanzanian Kilimanjaro Medal; The Primate Society of Great Britain Conservation Award; The Caring Institute Award; The Polar Bear Award; William Procter Prize for Scientific Achievement
- 1997: John & Alice Tyler Prize for Environmental Achievement; David S. Ingells, Jr. Award for Excellence; Common Wealth Award for Public Service; The Field Museum's Award of Merit; Tyler Prize for Environmental Achievement; Royal Geographical Society / Discovery Channel Europe Award for A Lifetime of Discovery
- 1998: Disney's Animal Kingdom Eco Hero Award; National Science Board Public Service Award; The Orion Society's John Hay Award
- 1999: International Peace Award; Botanical Research Institute of Texas International Award of Excellence in Conservation, Community of Christ International Peace Award

- 2001: Graham J. Norton Award for Achievement in Increasing Community Livability; Rungius Award of the National Museum of Wildlife Art, USA; Roger Tory Peterson Memorial Medal, Harvard Museum of Natural History; Master Peace Award; Gandhi/King Award for Non-Violence
- 2002: The Huxley Memorial Medal, Royal Anthropological Institute of Great Britain and Ireland; United Nations "Messenger of Peace" Appointment
- 2003: Benjamin Franklin Medal in Life Science; Harvard Medical School's Center for Health and the Global Environment Award; Prince of Asturias Award for Technical and Scientific Achievement; Dame Commander of the Order of the British Empire, presented by His Royal Highness Prince Charles; Chicago Academy of Sciences' Honorary Environmental Leader Award
- 2004: Nierenberg Prize for Science in the Public Interest; Will Rogers Spirit Award, the Rotary Club of Will Rogers and Will Rogers Memorial Museums; Life Time Achievement Award, the International Fund for Animal Welfare; Honorary Degree from Haverford College
- 2005: Honorary doctorate degree in science from Syracuse University
- 2005: Presented with Discovery and Imagination Award
- 2006: Received the 60th Anniversary Medal of the UNESCO and the French Légion d'honneur.
- 2007: Honorary doctorate degree in commemoration of Carl Linnaeus from Uppsala University
- 2007: Honorary doctorate degree from University of Liverpool
- 2008: Honorary doctorate degree from University of Toronto
- 2011: Honorary doctorate degree from American University of Paris

A complete list of Goodall's awards and honors is available through her curriculum vitae on the Jane Goodall Institute website.

67

Amy Johnson

Amy Johnson CBE, (1 July 1903 – 5 January 1941) was a pioneering English aviator. Flying solo or with her husband, Jim Mollison, Johnson set numerous long-distance records during the 1930s. Johnson flew in the Second World War as a part of the Air Transport Auxiliary where she died during a ferry flight.

Early Life

Johnson was born in Kingston upon Hull and was educated at Boulevard Municipal Secondary School (later Kingston High School). and the University of Sheffield, where she graduated with a Bachelor of Arts degree in economics. She then worked in London as secretary to the solicitor, William Charles Crocker. She was introduced to flying as a hobby, gaining a pilot's "A" Licence, No. 1979 on 6 July 1929 at the London Aeroplane Club under the tutelage of Captain Valentine Baker. In that same year, she became the first British woman to obtain a ground engineer's "C" licence.

Aviation Career

Her father, always one of her strongest supporters, offered to help her buy an aircraft. With funds from her father and Lord Wakefield she purchased *G-AAAH*, a second-hand de Havilland Gipsy Moth she named "Jason", not after the voyager of Greek legend, but after her father's trade mark. Johnson achieved worldwide recognition when, in 1930, she became the first woman to fly solo from Britain to Australia. Flying her "Jason" Gipsy Moth, she left Croydon, south of London, on 5 May of that year and landed in Darwin, Australia on 24 May after flying 11,000 miles (18,000 km). Her aircraft for this flight can still be seen in the Science Museum in London. She received the Harmon Trophy as well as a CBE in recognition of this achievement, and was also honoured with the No. 1 civil pilot's licence under Australia's 1921 Air Navigation Regulations. In July 1931, Johnson and her co-pilot Jack Humphreys, became the first pilots to fly from London to Moscow in one day, completing the 1,760 miles (2,830 km) journey in approximately 21 hours. From there, they continued across Siberia and on to Tokyo, setting a record time for flying from England to Japan. The flight was completed in a de Havilland Puss Moth. In 1932, Johnson married famous Scottish pilot Jim Mollison, who had, during a flight together, proposed to her only eight hours after they had met. In July 1932, Johnson set a solo record for the flight from London to Cape Town, South Africa in a Puss Moth, breaking her new husband's record. Her next flights were as a duo, flying with Mollison, she flew *G-ACCV* "Seafarer," a de Havilland Dragon Rapide nonstop from Pendine Sands, South Wales, to the United States in 1933. However, their aircraft ran out of fuel and crash-landed in Bridgeport, Connecticut; both were injured. After recuperating, the pair were feted by New York society and received a ticker tape parade down Wall Street. The Mollisons also flew in record time from Britain to India in 1934 in a de Havilland DH.88 Comet as part of the Britain to Australia MacRobertson Air Race. They were forced to retire from the race at Allahabad because of engine trouble. In May 1936, Johnson made her last record-breaking flight, regaining her Britain to South Africa record in *G-ADZO*, a Percival Gull Six. In 1938 Johnson divorced Mollison. Soon afterwards she reverted to her maiden name.

Second World War

In 1940, during the Second World War, Johnson joined the newly formed ATA, whose job was to transport Royal Air Force aircraft around the country – and rose to First Officer. (Her ex-husband Jim Mollison also flew for the ATA throughout the war.)

Death

On 5 January 1941, while flying an Airspeed Oxford for the Air Transport Auxiliary from Blackpool to RAF Kidlington near Oxford, Johnson went off course in adverse weather conditions. Reportedly out of fuel, she drowned after bailing out into the Thames Estuary. Although she was seen alive in the water, a rescue attempt failed and her body was never recovered. The incident also led to the death of her would-be rescuer, Lt Cmdr Walter Fletcher of HMS *Haslemere*. A memorial service was held in the church of St. Martin's in the Fields on 14 January 1941.

Disputed Circumstances

There is still some mystery about the accident, as the exact reason for the flight is still a government secret and there is some evidence that besides Johnson and Fletcher a third person (possibly someone she was supposed to ferry somewhere) was also seen in the water and also drowned. Who the third party was is still unknown. Johnson was the first member of the Air Transport Auxiliary to die in service. Her death in an Oxford aircraft was ironic, as she had been one of the original subscribers to the share offer for Airspeed. However, in 1999 it was reported that Tom Mitchell, from Crowborough, Sussex, claimed to have shot the heroine down when she twice failed to give the correct identification code during the flight. He said: "The reason Amy was shot down was because she gave the wrong colour of the day [a signal to identify aircraft known by all British forces] over radio." Mr. Mitchell explained how the aircraft was sighted and contacted by radio. A request was made for the signal. She gave the wrong one twice. "Sixteen rounds of shells were fired and the plane dived into the Thames Estuary. We all thought it was an enemy plane until the next day when we read the papers and discovered it was Amy. The officers told us never to tell anyone what happened."

Honours and Tributes

During her life, Johnson was recognised in many ways. In June 1930, Johnson's flight to Australia was the subject of a contemporary popular song, "Amy, Wonderful Amy", composed by Horatio Nicholls and recorded by Harry Bidgood, Jack Hylton, Arthur Lally, Arthur Rosebery and Debroy Somers. She was also the guest of honour at the opening of the first Butlins holiday camp, in Skegness in 1936. From 1935 to

1937, Johnson was the President of the Women's Engineering Society. A collection of Amy Johnson souvenirs and mementos was donated by her father to Sewerby Hall in 1958. The hall now houses a room dedicated to Amy Johnson in its museum. In 1974, Harry Ibbetson's statue of Amy Johnson was unveiled in Prospect Street, Kingston upon Hull where a girls' school was named after her (the school later closed in 2004). Public edifices to Johnson's honour includes the "Amy Johnson Building" housing the department of Automatic Control and Systems Engineering at the University of Sheffield is named after her. The "Amy Johnson Primary School" situated on Mollison Drive on the Roundshaw Estate, Wallington, Surrey, is named after Johnson and built on the former runway site of Croydon Airport. Street names named in her honour include

- "Amy Johnson Avenue", a major arterial road in Darwin, Australia connecting the Stuart Highway to Old McMillan's Road
- "Amy Johnson Avenue" in Bridlington, East Riding of Yorkshire
- "Amy Johnson Way", close to Blackpool Airport, in Blackpool, Lancashire
- "Amy Johnson Way" in the Rawcliffe area of York
- "Mollison Way" in Queensbury, London.

Other tributes to Johnson include a KLM McDonnell-Douglas MD-11 named in her honour and "Amy's Restaurant and Bar" at the Hilton Stansted, London named after her. In 2011 the Royal Aeronautical Society established the annual Amy Johnson Named Lecture to celebrate a century of women in flight and to honour Britain's most famous woman aviator. Carolyn McCall, Chief Executive of EasyJet, will deliver the Inaugural Lecture on the 6 July 2011 at the Society's headquarters in London. The Lecture will be held on or close the 6 July every year to mark the date in 1929 when Amy Johnson was awarded her pilot's licence.

68

Martina Navratilova

Martina Navratilova (born October 18, 1956) is a Czech American tennis player and a former World No. 1. Billie Jean King said about Navratilova in 2006, "She's the greatest singles, doubles and mixed doubles player who's ever lived." Navratilova won 18 Grand Slam singles titles, 31 Grand Slam women's doubles titles (an all-time record), and 10 Grand Slam mixed doubles titles. She reached the Wimbledon singles final 12 times, including 9 consecutive years from 1982 through 1990, and won the women's singles title at Wimbledon a record 9 times. She and King each won 20 Wimbledon titles, an all-time record. Navratilova is one of just three women to have accomplished a career Grand Slam in singles, women's doubles, and mixed doubles (called the Grand Slam "boxed set") a record she shares with Margaret Court and Doris Hart. She holds the open era record for most singles titles (167) and doubles titles (177). She recorded the longest winning streak in the open era (74 consecutive matches) and three of the six longest winning streaks in the women's open era. Navratilova, Margaret Court, and Maureen Connolly share the record for the most consecutive Grand Slam singles titles (six). Navratilova reached 11 consecutive Grand Slam singles finals, second all-time to Steffi Graf's 13.

In women's doubles, Navratilova and Pam Shriver won 109 consecutive matches and won all four Grand Slam titles in 1984. Also the pair set an all time record of 79 titles together and tied Louise Brough Clapp's and Margaret Osborne duPont's record of 20 Grand Slam women's doubles titles as a team. In addition she won the season ending WTA Tour Championships a record 8 times and made the finals a record

14 times. Navratilova is the only man or woman to have won 8 different tournaments at least 7 times. Originally from Czechoslovakia, she was stripped of her citizenship when, in 1975 at the age of 18, she asked the United States for political asylum and was granted temporary residency. At the time, Navratilova was told by the Czechoslovakian Sports Federation that she was becoming too Americanized and that she should go back to school and make tennis secondary. Navratilova became a US citizen in 1981, but on January 9, 2008, she had her Czech citizenship restored. She stated she has not renounced her American citizenship nor does she plan to do so and that the restoration of her Czech citizenship was not politically motivated. Navratilova is a member of the Laureus World Sports Academy. She also serves the Health and Fitness Ambassador for AARP in an alliance created to help AARP's millions of members lead active, healthy lives.

Early Life and Tennis Career

Navratilova was born Martina Šubertová in Prague, Czechoslovakia. Her parents divorced when she was three, and in 1962 her mother Jana married Miroslav Navrátil, who became her first tennis coach. Martina then took the name of her stepfather. Her father Mirek remarried and divorced. When she was eight, he committed suicide. In 2008, Navratilova's mother died of emphysema, aged 75. Navratilova has a sister, Jana, and an older paternal half-brother. In 1972 at the age of 15, Navratilova won the Czechoslovakia national tennis championship. In 1973, aged 16, she made her debut on the United States Lawn Tennis Association professional tour but did not turn professional until 1975. She won her first professional singles title in Orlando, Florida in 1974 at the age of 17. Upon arriving in the United States, Navratilova first lived with former Vaudeville actress, Frances Dewey Wormser, and her husband, Morton Wormser, a tennis enthusiast. Navratilova was the runner-up at two Grand Slam singles tournaments in 1975. She lost in the final of the Australian Open to Evonne Goolagong Cawley and in the final of the French Open to Chris Evert. After losing to Evert in the semifinals of that year's US Open, the 18-year-old Navratilova went to the offices of the Immigration and

Naturalization Service in New York City and informed them that she wished to defect from Communist Czechoslovakia. Within a month, she received a green card. Navratilova won her first Grand Slam singles title at Wimbledon in 1978, where she defeated Evert in three sets in the final and captured the World No. 1 ranking for the first time. She successfully defended her Wimbledon title in 1979, again beating Evert in the final, and retained her World No. 1 ranking. In 1981, Navratilova won her third Grand Slam singles title by defeating Evert in the final of the Australian Open. Navratilova also reached the final of the US Open, where she lost a third set tiebreak to Tracy Austin. Navratilova won both Wimbledon and the French Open in 1982. After adopting basketball player Nancy Lieberman's exercise plan and using graphite racquets, Navratilova became the most dominant player in women's tennis. After losing in the fourth round of the first Grand Slam event of 1983, the French Open, she captured the year's three remaining Grand Slam titles (the Australian Open was held in December at that time). Navratilova's loss at the French Open was her only singles defeat during that year, during which she established an 86–1 record. Her winning percentage was the best ever for a post-1968 professional tennis player. During 1982, 1983, and 1984, Navratilova lost a total of only six singles matches.

Navratilova won the 1984 French Open, thus holding all four Grand Slam singles titles simultaneously. Her accomplishment was declared a "Grand Slam" by Philippe Chatrier, president of the International Tennis Federation. Many tennis observers, however, insisted that it was not a true Grand Slam because the titles had not been won in a single calendar year. Navratilova extended her Grand Slam singles tournament winning streak to a record-equalling six following wins at Wimbledon and the US Open. She entered the 1984 Australian Open with a chance of winning all four titles in the same year. In the semifinals, however, Helena Suková ended Navratilova's 74-match winning streak (a record for a professional) 1–6, 6–3, 7–5. A left-hander, Navratilova won all four Grand Slam women's doubles titles in 1984, partnering right-handed Pam Shriver, a tall and talented player whose most noted stroke was a slice forehand, a shot virtually unheard of in the game today. This was part of a record 109-match winning streak that the pair achieved between 1983 and 1985. (Navratilova was ranked the World No. 1 doubles player for a period of over three years in the 1980s.) From 1985 through 1987, Navratilova reached the women's singles final at all 11 Grand Slam tournaments held during those three years, winning six of them. From 1982 through 1990, she reached the Wimbledon final nine consecutive times. She reached the US Open final five consecutive times from 1983 through 1987 and appeared in the French Open final five out of six years from 1982 through 1987.

17-year old German player Steffi Graf emerged on the scene in 1987 when she beat Navratilova in the final of the French Open. Navratilova defeated Graf in the 1987 Wimbledon and US Open finals (and at the US Open became only the third player in the open era to win the women's singles, women's doubles, and mixed doubles at the same event). Graf's consistent play throughout 1987, however, allowed her to obtain the World No. 1 ranking before the end of the year. Graf eventually broke Navratilova's records of 156 consecutive weeks and 331 total weeks as the World No. 1 singles player but did not break Navratilova's record 167 singles titles as Graf reached 107. In 1988, Graf won all four Grand Slam singles titles, beating Navratilova 5–7, 6–2, 6–1 in the Wimbledon final along the way. In 1989, Graf and Navratilova met in the finals of both Wimbledon and the US Open, with Graf winning both encounters in three sets. Despite the age difference between the two players, Navratilova won 9 of the 18 career singles matches with Graf and 5 of the 9 Grand Slam singles matches with her. At age 34, Navratilova defeated Graf the last time they played in a Grand Slam event in the semifinals of the 1991 US Open 7–6(2), 6–7(6), 6–4.

Personal Life

In 1981, shortly after becoming a United States citizen, Navratilova came out publicly about her sexual orientation. During the early 1980s, she was involved with author Rita Mae Brown. From 1984 to 1991, Navratilova had a long-term relationship with partner Judy Nelson. Their split in 1991 included a much-publicized legal wrangle. Navratilova was featured in a WITA (Women's International Tennis Association) calendar, shot by Jean Renard with her Wimbledon trophies and Nelson's children in the background. In 1985, Navratilova released an autobiography, co-written with *New York Times* sports columnist George Vecsey, entitled *Martina* in the U.S. and *Being Myself* in the rest of the world. She had earlier co-written a tennis instruction book with Mary Carillo in 1982 entitled *Tennis My Way*. She later wrote three mystery novels with Liz Nickles: *The Total Zone* (1994), *Breaking Point* (1996), and *Killer Instinct* (1997). Navratilova's most recent literary effort was a health and fitness book entitled *Shape Your Self,* which came out in 2006. On April 7, 2010, Navratilova announced that she was being treated for breast cancer. A routine mammography in January 2010 had revealed that she had a ductal carcinoma in situ in her left breast, which she was informed of in February, and in March she had the tumour surgically removed; she received radiation therapy in May. In December 2010, Navratilova was hospitalized after developing high altitude pulmonary edema while attempting a climb of Mt. Kilimanjaro in Tanzania.

Awards

- ITF World Champion 1979, 1982, 1983, 1984, 1985, 1986.
- WTA Player of the Year 1978, 1979, 1982, 1983, 1984, 1985, 1986.

Recognition

Tennis magazine has selected her as the greatest female tennis player for the years 1965 through 2005. Tennis historian and journalist Bud Collins has called Navratilova "arguably, the greatest player of all time." Tennis writer Steve Flink, in his book *The Greatest Tennis Matches of the Twentieth Century*, named her as the second best female player of the 20th century, directly behind Steffi Graf.

I'm a Celebrity

In November 2008, Martina Navratilova appeared on the UK's ITV series Series 8 of I'm a Celebrity... Get Me Out of Here!; she finished runner-up to Joe Swash.

69

Daphne du Maurier

Dame Daphne du Maurier, Lady Browning (13 May 1907 – 19 April 1989) was an English author and playwright. Many of her works have been adapted into films, including the novels *Rebecca* (which won the Best Picture Oscar in 1941) and *Jamaica Inn* and the short stories "The Birds" and "Don't Look Now". The first three were directed by Alfred Hitchcock. Her elder sister was the writer Angela du Maurier. Her father was the actor Gerald du Maurier. Her grandfather was the writer George du Maurier.

Early Life

Daphne du Maurier was born in London, the second of three daughters of the prominent actor-manager Sir Gerald du Maurier and actress Muriel Beaumont (maternal niece of William Comyns Beaumont). Her grandfather was the author and *Punch* cartoonist George du Maurier, who created the character of Svengali in the novel *Trilby*. These connections helped her in establishing her literary career, and du Maurier published some of her very early work in Beaumont's *Bystander* magazine. Her first novel, *The Loving Spirit,* was published in 1931. Du Maurier was also the cousin of the Llewelyn Davies boys, who served as J.M. Barrie's inspiration for the characters in the play *Peter Pan, or The Boy Who Wouldn't Grow Up*. As a young child, she met many of the brightest stars of the theatre, thanks to the celebrity of her father. On meeting Tallulah Bankhead, she was quoted as saying that the actress was the most beautiful creature she had ever seen.

Novels, Short Stories and Biographies

Literary critics have sometimes berated du Maurier's works for not being "intellectually heavyweight" like those of George Eliot or Iris Murdoch. By the 1950s, when the socially and politically critical "angry young men" were in vogue, her writing was felt by some to belong to a bygone age. Today, she has been reappraised as a first-rate storyteller, a mistress of suspense. Her ability to recreate a sense of place is much admired, and her work remains popular worldwide. For several decades she was the most popular author for library book borrowings. The novel *Rebecca*, which has been adapted for stage and screen on several occasions, is generally regarded as her masterpiece. One of her strongest influences here was *Jane Eyre* by Charlotte Brontë. Her fascination with the Brontë family is also apparent in *The Infernal World of Branwell Brontë*, her biography of the troubled elder brother to the Brontë girls. The fact that their mother had been Cornish no doubt added to her interest. Other notable works include *The Scapegoat*, *The House on the Strand*, and *The King's General*. The latter is set in the middle of the first and second English Civil Wars. Though written from the Royalist perspective of her native Cornwall, it gives a fairly neutral view of this period of history.

Du Maurier's novel *Mary Anne* (1954) is a fictionalised account of the real-life story of her great-great-grandmother, Mary Anne Clarke née Thompson (1776–1852). From 1803 to 1808, Mary Anne Clarke was mistress of Frederick Augustus, Duke of York and Albany (1763–1827). He was the "Grand Old Duke of York" of the nursery rhyme, a son of King George III and brother of the later King George IV. In Ken Follett's thriller *The Key to Rebecca*, du Maurier's novel *Rebecca* is used as the key for a code used by a German spy in World War II Cairo. Neville Chamberlain is reputed to have read *Rebecca* on the plane journey that led to Adolf Hitler signing the Munich Agreement. The central character of her last novel, *Rule Britannia*, is an aging and eccentric actress who was based on Gertrude Lawrence and Gladys Cooper (to whom it is dedicated). However, the character is most recognisably du Maurier herself. Indeed, it was in her short stories that she was able to give free rein to the harrowing and terrifying side of her imagination; "The Birds", "Don't Look Now", "The Apple Tree" and "The Blue Lenses" are exquisitely crafted tales of terror that shocked and surprised her audience in equal measure. A more recent discovery of a collection of du Maurier's forgotten short stories, written when the author was 21, provides an intriguing insight into the writer she was to become. One of them, "The Doll", is a suspense-driven gothic tale about a young woman's obsession with a mechanical male sex doll; it has been deemed by du Maurier's son Kit Browning as being "quite ahead of its time".

Perhaps more than at any other time, du Maurier was anxious as to how her bold new writing style would be received, not just by her readers (and to some extent her critics, though by then she had grown wearily accustomed to their often lukewarm reviews) but also by her immediate circle of family and friends. In later life, she wrote nonfiction, including several biographies that were well received. This, no doubt, came from a deep-rooted desire to be accepted as a serious writer, comparing herself to her neighbour, A. L. Rowse, the celebrated historian and essayist, who lived a few miles away from her house near Fowey. Also of interest are the "family" novels/ biographies that du Maurier wrote of her own ancestry, of which *Gerald*, the biography of her father, was most lauded. Later she wrote *The Glass-Blowers*, which traces her French ancestry and gives a vivid depiction of the French Revolution. *The du Mauriers* is a sequel of sorts describing the somewhat problematic ways in which the family moved from France to England in the 19th century and finally *Mary Anne*, the novel based on the life of a notable, and infamous, English ancestor—her great-grandmother Mary Anne Clarke, former mistress of Frederick, Duke of York.

Her final novels reveal just how far her writing style had developed. *The House on the Strand* (1969) combines elements of "mental time-travel", a tragic love affair in 14th century Cornwall, and the dangers of using mind-altering drugs. Her final novel, *Rule Britannia*, written post-Vietnam, plays with the resentment of English people in general and Cornish people in particular at the increasing dominance of the U.S.

In late 2006, a previously unknown work titled *And His Letters Grew Colder* was discovered by Ann Willmore of Bookends of Fowey. This was estimated to have been written in the late 1920s and takes the form of a series of letters tracing an adulterous, passionate affair from initial ardour to deflated acrimony.

Plays

Daphne du Maurier wrote three plays. Her first was a successful adaptation of her novel *Rebecca*, which opened at the Queen's Theatre in London on 5 March 1940 in a production by George Devine, starring Celia Johnson and Owen Nares as the De Winters and Margaret Rutherford as Mrs. Danvers. At the end of May, following a run of 181 performances, the production transferred to the Strand Theatre, with Jill Furse taking over as Mrs. De Winter and Mary Merrall as Danvers, with a further run of 176 performances. In the summer of 1943, she began writing the autobiographically inspired drama *The Years Between* about the unexpected return of a senior officer, thought killed in action, who finds that his wife has taken his seat as Member of Parliament and has started a romantic relationship with a local farmer. It was first staged at the Manchester Opera House in 1944 and then transferred to London, opening at Wyndham's Theatre on 10 January 1945, starring Nora Swinburne and Clive Brook. The production, directed by Irene Hentschel, became a long-running hit, completing 617 performances. After 60 years of neglect, it was revived by Caroline Smith at the Orange Tree Theatre in Richmond upon Thames on 5 September 2007, starring Karen Ascoe and Mark Tandy.

Personal Names, Titles and Honours

She was known as Daphne du Maurier from 1907 to 1932 when she became Mrs Frederick Browning while writing as Daphne du Maurier (1932–1946). She was titled Lady Browning; Daphne du Maurier (1946–1969). Later, on receiving the honorific Dame Commander of the Order of the British Empire, she was Lady Browning; Dame Daphne du Maurier DBE (1969–1989). When in the Queen's Birthday Honours List for June 1969 Daphne du Maurier was created a Dame Commander of the Order of the British Empire, she accepted but never used the title. According to Margaret Forster, she told no one about the honour, so that even her children only learned of it from the newspapers. "She thought of pleading illness for the investiture, until her children insisted it would be a great day for the older grandchildren. So she went through with it, though she slipped out quietly afterwards to avoid the attention of the press".

Personal Life

She married Lieutenant-General Sir Frederick "Boy" Browning, with whom she had two daughters, Tessa and Flavia, and a son, Christian. Tessa Browning (b. 1933) married Major Peter de Zulueta, whom she divorced and married David Montgomery, 2nd Viscount Montgomery of Alamein in 1970. Flavia (b. 1937) married Captain Alastair Tower, whom she divorced, before marrying General Sir Peter Leng. Christian (b. 1940) became a photographer and film-maker. Biographers have noted that the marriage was at times somewhat chilly and that du Maurier could be aloof and distant to her children, especially the girls, when immersed in her writing. "Boy" died in 1965 and soon after Daphne moved to Kilmarth, near Par, which became the setting for *The House on the Strand*. Du Maurier has often been painted as a frostily private recluse who rarely mixed in society or gave interviews. An exception to this came after the release of the film *A Bridge Too Far*, in which her late husband was portrayed in a less-than-flattering light. Du Maurier, incensed, wrote to the national newspapers, decrying what she considered unforgivable treatment. Once out of the glare of the public spotlight, however, many remembered her as a warm and immensely funny person who was a welcoming hostess to guests at Menabilly, the house she leased for many years (from the Rashleigh family) in Cornwall.

Death

Du Maurier died aged 81 at her home in Cornwall, which had been the setting for many of her books. Her body was cremated and her ashes scattered at Kilmarth.

70

Golda Meir

Golda Meir (3 May 1898 – 8 December 1978) was a teacher, *kibbutznik* and politician who became the fourth Prime Minister of the State of Israel. Meir was elected Prime Minister of Israel on 17 March 1969, after serving as Minister of Labour and Foreign Minister. Israel's first and the world's third woman to hold such an office, she was described as the "Iron Lady" of Israeli politics years before the epithet became associated with British prime minister Margaret Thatcher. Former prime minister David Ben-Gurion used to call Meir "the best man in the government"; she was often portrayed as the "strong-willed, straight-talking, grey-bunned grandmother of the Jewish people."

Early Life

Golda Mabovitch was born on 3 May, 1898 in Kiev, Russian Empire, in present day Ukraine, to Blume Neiditch and Moshe Mabovitch, a carpenter. Meir wrote in her autobiography that her earliest memories were of her father boarding up the front door in response to rumors of an imminent pogrom. She had two sisters, Sheyna and Tzipke, as well as five other siblings who died in childhood. She was especially close to Sheyna. Moshe Mabovitch left to find work in New York City in 1903. In his absence, the rest of the family moved to Pinsk to join her mother's family. In 1905, Moshe moved to Milwaukee, WI in search of higher-paying work and found employment in the workshops of the local railroad yard. The following year, he had saved up enough money to bring his family to the United States. Blume ran a grocery store on Milwaukee's north side, where by age eight Golda had been put in charge of watching the store when her mother went to the market for supplies. Golda attended the Fourth Street Grade School (now Golda Meir School) from 1906 to 1912. A leader early on, she organized a fund raiser to pay for her classmates' textbooks. After forming the American Young Sisters Society, she rented a hall and scheduled a public meeting for the event. She went on to graduate valedictorian of her class, despite not knowing English at the beginning of her schooling. At 14, she studied at North Division High School and worked part-time. Her mother wanted her to leave school and marry, but she rebelled. She bought a train ticket to Denver, Colorado, and went to live with her married sister, Sheyna Korngold. The Korngolds held intellectual evenings at their home, where Meir was exposed to debates on Zionism, literature, women's suffrage, trade unionism, and more. In her autobiography, she wrote: "To the extent that my own future convictions were shaped and given form... those talk-filled nights in Denver played a considerable role." In Denver, she also met Morris Meyerson, a sign painter, whom she later married on 24 December, 1917.

Return to Milwaukee, Zionist Activism and Teaching

In 1913, she returned to North Division High, graduating in 1915. While there, she became an active member of Young Poale Zion, which later became Habonim, the Labor Zionist youth movement. She spoke at public meetings, embraced Socialist Zionism and hosted visitors from Palestine. She attended the teachers college, Milwaukee State Normal School, (now University of Wisconsin–Milwaukee) in 1916, and probably part of 1917. After graduating from Milwaukee State, she taught in Milwaukee Public Schools (MPS). In 1917, she took a position at a Yiddish-speaking Folks Schule in Milwaukee. While at the Folks Schule, she came more closely into contact with the ideals of Labor Zionism. In 1913, she began dating Morris Meyerson. She was a committed Labor Zionist and he was a dedicated socialist. Together, they left their jobs to join a kibbutz in Palestine in 1921. When Golda and Morris married in 1918, settling in Palestine was her precondition for the marriage. Golda had intended to make Aliyah straight away but her plans were disrupted due to all transatlantic passenger services being canceled due to the first world war. Instead she threw her energies into Poale Zion activities. A short time after their wedding, she embarked on a fund raising campaign for Poale Zion that took her across the United States.The couple moved to Palestine in 1921 together with her sister Shayna.

Pre-state Political Role

In June 1946, the British cracked down on the Zionist movement in Palestine, arresting many leaders of the Yishuv. They had been provoked by paramilitary Zionist activities. Meir took over as acting head of the Political Department of the Jewish Agency during the incarceration of Moshe Sharett. Thus she became the principal negotiator between the Jews in Palestine and the British Mandatory authorities. After his release, Sharett went to the United States to attend talks on the UN Partition Plan, leaving Meir to head the Political Department until the establishment of the state in 1948. In January 1948, the treasurer of the Jewish Agency was convinced that Israel would not be able to raise more than $7–8 million from the American Jewish community. Meir traveled to the United States and managed to raise $50 million, which was used to purchase arms in Europe for the nascent state. Ben-Gurion wrote that Meir's role as the "Jewish woman who got the money which made the state possible" would go down one day in the history books. On 10 May 1948, four days before the official establishment of the state, Meir traveled to Amman disguised as an Arab woman for a secret meeting with King Abdullah of Transjordan at which she urged him not to join the other Arab countries in attacking the Jews. Abdullah asked her not to hurry to proclaim a state. Meir replied: "We've been waiting for 2,000 years. Is that hurrying?" As head of the Jewish Agency Political Department, Meir called the mass exodus of Arabs before the War of Independence in 1948 as "dreadful" and likened it to what had befallen the Jews in Nazi-occupied Europe.

Ministerial Career

Meir was one of twenty-four signatories (two of them women) of the Israeli declaration of independence on 14 May 1948. She later recalled, "After I signed, I cried. When I studied American history as a schoolgirl and I read about those who signed the Declaration of Independence, I couldn't imagine these were real people doing something real. And there I was sitting down and signing a declaration of establishment." Israel was attacked the next day by the joint armies of Egypt, Syria, Lebanon, Transjordan, and Iraq in the 1948 Arab-Israeli War.

Ambassador to Moscow

Carrying the first Israeli-issued passport, Meir was appointed Israel's ambassador to the Soviet Union. During her brief stint there, which ended in 1949, she attended high holiday services at the synagogue in Moscow, where she was mobbed by thousands of Russian Jews chanting her name. The Israeli 10,000 shekel banknote issued in November 1984 bore a portrait of Meir on one side and the image of the crowd that turned out to cheer her in Moscow on the other.

Labour Minister

In 1949, Meir was elected to the Knesset as a member of Mapai and served continuously until 1974. From 1949 to 1956, she served as Minister of Labour, introducing major housing and road construction projects. In 1955, on Ben Gurion's instructions, she stood for the position of mayor of Tel Aviv. She lost by the two votes of the religious bloc who withheld their support on the grounds that she was a woman.

Death

On 8 December, 1978, Meir died of lymphatic cancer in Jerusalem at the age of 80. Four days later, on 12 December, Meir was buried on Mount Herzl in Jerusalem.

Awards

In 1975, Meir was awarded the Israel Prize for her special contribution to society and the State of Israel. In 1974, Meir was awarded the honor of World Mother by American Mothers, Inc.

71

Tegla Loroupe

Tegla Chepkite Loroupe (born 9 May 1973 in Kutomwony, Kenya) is a long-distance track and road runner, and a global spokeswoman for peace, women's rights, and education. Loroupe holds the world records for 20, 25 and 30 kilometres and previously held the world marathon record. She is the three-time World Half-Marathon champion. She was the first African woman to win the New York City Marathon, which she has won twice. She has won marathons in London, Boston, Rotterdam, Hong Kong, Berlin, Rome and many of other cities.

Career

See the IAAF (International Athletic Federation) official biography for more detail. Tegla Loroupe was born in *Kutomwony* village, *Lelan* division of West Pokot District, in the Rift Valley approximately 600 kilometers north of Nairobi. Her father and mother are of the Pokot tribe, who live in the area of northern Kenya, eastern Uganda and southern Ethiopia. She grew up with 24 siblings; the Pokots are a polygamous culture; her father had four wives. She spent her childhood working fields, tending cattle and looking after younger brothers and sisters. At the age of six she started school at Kapsait Elementary school; she had to run ten kilometres to and from school every morning. At school, she became aware of her potential as an athlete when she won races held over a distance of 800 or 1500 metres against students much older again. She decided to pursue a career as a runner, but - except for her mother - was not supported by anyone. The Kenyan athletics federation, Athletics Kenya, did not support her at first, thinking

Loroupe too small and too thin. However, after she won a prestigious cross country barefoot race in 1988, this changed. She began to train to compete internationally the following year, earning her first pair of running shoes in 1989, which she wore only for particularly rough races. She was nominated for the junior race of the 1989 IAAF World Cross Country Championships finishing 28th. She competed again at the 1990 IAAF World Cross Country Championships, finishing 16th in the junior race.

In 1994 and 1998, Loroupe won the Goodwill Games over 10,000 metres, barefoot. Over the same distance she won bronze medals at the IAAF World Championships in Athletics 1995 and 1999. In 1994, Loroupe ran her first major marathon in New York. Running against the world's strongest competition, she won. As a consequence she was idolized by many young people in Africa: at last, a woman champion to complement the many successful male runners. She won the New York City Marathon again in 1995 and finished 3rd in 1998. Between 1997 and 1999, she won three world titles a row over the half marathon distance. She won Rotterdam Marathon three times between 1997 and 1999. She won Berlin Marathon in 1999 and finished second in 2001. She finished second at the 1999 Osaka International Ladies Marathon. Loroupe won the Zevenheuvelenloop 15K race in the Netherland three times (1992, 1993 and 1998). She is a seven-time Egmond Half Marathon winner (1993–1998, 2000). She has won the Lisbon Half Marathon a record six times: 1994-1997, 1999 and 2000. She has won the Tilburg road race, a five times (1993, 1994, 1996, 1998, 1999), also a record number. She won the Paris Half Marathon in 1994 and 1998, City-Pier-City Loop half marathon in the Hague in 1998, and the Parelloop 10K in race in the Netherlands in 1999

During the 2000 Summer Olympics in Sydney, Australia, favored to win both the marathon and the 10,000 meters, she suffered from violent food poisoning the night before the race. Nevertheless she fought through the marathon race, finishing 13th, then, the next day, ran the 10,000 metres, finishing 5th, running barefoot in both races, a feat she later stated she achieved out of a sense of duty to all the people taking her as a bearer of hope in her home country. Until the end of 2001, she continued to suffer from various health problems.

Goodwill

In 2006, she was named a United Nations Ambassador of Sport by Secretary General Kofi Annan, together with Roger Federer, tennis champion from Switzerland, Elias Figueroa, Latin American soccer legend from Chile, and Katrina Webb paralympics gold medalist from Australia. She is an International Sports Ambassador for the IAAF, the International Association of Athletics Federations, and for UNICEF. In 2003, Loroupe created an annual series of Peace Marathons sponsored by the Tegla Loroupe Peace Foundation "Peace Through Sports". Presidents, Prime Ministers, Ambassadors and government officials run with warriors and nomadic groups in her native Kenya, in Uganda and in Sudan, to bring peace to an area plagued by raiding warriors from battling tribes. In 2010 the Kenyan Government lauded her achievements as hundreds of warriors had laid down their weapons. She has established a school (Tegla Loroupe Peace Academy) and orphanage for children from the region in Kapenguria, a high-mountain town in north-west Kenya. The 2006 Peace Marathon was held on 18 November 2006, in Kapenguria, Kenya. Two thousand warriors from six tribes competed. The next Peace Marathon is 15 November 2008 in Kapenguria, Kenya. Many ambassadors to Kenya are expected, together with the Prime Minister, and several Kenyan, Tanzanian, and Ugandan Ministers. In February, 2007, she was named the Oxfam Ambassador of Sport and Peace to Darfur. In December, 2006, she traveled with George Clooney, Joey Cheek, and Don Cheadle to Beijing, Cairo, and New York on a diplomatic mission to bring an end to violence in Darfur. She won the "Community Hero" category at the 2007 Kenyan Sports Personality of the Year awards. Tegla Loroupe is today a member of the 'Champions for Peace' club, a group of 54 famous elite athletes committed to serving peace in the world through sport, created by Peace and Sport, a Monaco-based international organization.

72

Elizabeth Taylor

Dame Elizabeth Rosemond "Liz" Taylor, (February 27, 1932 – March 23, 2011) was a British-American actress. From her early years as a child star with MGM, she became one of the great screen actresses of Hollywood's Golden Age. As one of the world's most famous film stars, Taylor was recognized for her acting ability and for her glamorous lifestyle, beauty and distinctive violet eyes. *National Velvet* (1944) was Taylor's first success, and she starred in *Father of the Bride* (1950), *A Place in the Sun* (1951), *Giant* (1956), *Cat on a Hot Tin Roof* (1958), and *Suddenly, Last Summer* (1959). She won the Academy Award for Best Actress for *BUtterfield 8* (1960), played the title role in *Cleopatra* (1963), and married her co-star Richard Burton. They appeared together in 11 films, including *Who's Afraid of Virginia Woolf?* (1966), for which Taylor won a second Academy Award. From the mid-1970s, she appeared less frequently in film, and made occasional appearances in television and theatre. Her much publicized personal life included eight marriages and several life-threatening illnesses. From the mid-1980s, Taylor championed HIV and AIDS programs; she co-founded the American Foundation for AIDS Research in 1985, and the Elizabeth Taylor AIDS Foundation in 1993. She received the Presidential Citizens Medal, the Legion of Honour, the Jean Hersholt Humanitarian Award and a Life Achievement Award from the American Film Institute, who named her seventh on their list of the "Greatest American Screen Legends". Taylor died of congestive heart failure at the age of 79.

Early Years

Elizabeth Rosemond Taylor was born at Heathwood, her

parents' home at 8 Wildwood Road in Hampstead Garden Suburb, a northwestern suburb of London; the younger of two children of Francis Lenn Taylor (1897–1968) and Sara Viola Warmbrodt (1895–1994), who were Americans residing in England. Taylor's older brother, Howard Taylor, was born in 1929. Her parents were originally from Arkansas City, Kansas. Francis Taylor was an art dealer, and Sara was a former actress whose stage name was "Sara Sothern". Sothern retired from the stage in 1926 when she married Francis in New York City. Taylor's two first names are in honor of her paternal grandmother, Elizabeth Mary (Rosemond) Taylor. Colonel Victor Cazalet, one of their closest friends, had an important influence on the family. He was a rich, well-connected bachelor, a Member of Parliament and close friend of Winston Churchill. Cazalet loved both art and theater and was passionate when encouraging the Taylor family to think of England as their permanent home. Additionally, as a Christian Scientist and lay preacher, his links with the family were spiritual. He also became Elizabeth's godfather. In one instance, when she was suffering with a severe infection as a child, she was kept in her bed for weeks. She "begged" for his company: "Mother, please call Victor and ask him to come and sit with me." Biographer Alexander Walker suggests that Elizabeth's conversion to Judaism at the age of 27 and her life-long support for Israel, may have been influenced by views she heard at home.

A dual citizen of the United Kingdom and the United States, she was born a British subject through her birth on British soil and an American citizen through her parents. She reportedly sought, in 1965, to renounce her United States citizenship, to wit: "Though never accepted by the State Department, Elizabeth renounced in 1965. Attempting to shield much of her European income from U.S. taxes, Elizabeth wished to become solely a British citizen. According to news reports at the time, officials denied her request when she failed to complete the renunciation oath, refusing to say that she renounced "all allegiance to the United States of America." At the age of three, Taylor began taking ballet lessons. Shortly before the beginning of World War II, her parents decided to return to the United States to avoid hostilities. Her mother took the children first, arriving in New York in April 1939, while her father remained in London to wrap up matters in his art business, arriving in November. They settled in Los Angeles, California, where her father established a new art gallery, which included many paintings he shipped from England. The gallery would soon attract numerous Hollywood celebrities who appreciated its modern European paintings. According to Walker, the gallery "opened many doors for the Taylors, leading them directly into the society of money and prestige" within Hollywood's movie colony.

Acting Career

Soon after settling in Los Angeles, Taylor's mother discovered that Hollywood people "habitually saw a movie future for every pretty face." Some of her mother's friends, and even total strangers, urged her to have Taylor screen tested for the role of Bonnie Blue, Scarlett's child in *Gone with the Wind*, then being filmed. Her mother refused the idea, as a child actress in film was alien to her. And in any regard, they would return to England after the war. Hollywood columnist Hedda Hopper introduced the Taylors to Andrea Berens, the fiancée of Cheever Cowden, chairman and major stockholder of Universal Pictures.

The teenage Taylor was reluctant to continue making films. Her stage mother forced Taylor to relentlessly practice until she could cry on cue and watched her during filming, signaling to change her delivery or a mistake. Taylor met few others her age on movie sets, and was so poorly educated that she needed to use her fingers to do basic arithmetic. When at age 16 Taylor told her parents that she wanted to quit acting for a normal childhood, however, Sara Taylor told her that she was ungrateful: "You have a responsibility, Elizabeth. Not just to this family, but to the country now, the whole world". In October 1948, Taylor sailed aboard the RMS *Queen Mary* to England to begin filming *Conspirator*. Unlike some other child actors, Taylor made an easy transition to adult roles. Before *Conspirator*'s 1949 release, a *TIME* cover article called her "a jewel of great price, a true star sapphire", and the leader among Hollywood's next generation of stars such as Montgomery Clift, Kirk Douglas, and Ava Gardner.

Activist and Humanitarian

Taylor devoted consistent and generous humanitarian time, advocacy efforts, and funding to HIV and AIDS-related projects and charities, helping to raise more than $270 million for the cause. She was one of the first celebrities and public personalities to do so at a time when few acknowledged the disease, organizing and hosting the first AIDS fundraiser in 1984, to benefit AIDS Project Los Angeles. Taylor was cofounder of the American Foundation for AIDS Research (amfAR) with Dr. Michael Gottlieb and Dr. Mathilde Krim in 1985. Her longtime friend and former co-star Rock Hudson had disclosed having AIDS and died of it that year. After her conversion to Judaism, Taylor worked for Jewish causes throughout her life. In 1959, her large-scale purchase of Israeli Bonds triggered Arab boycotts of her films. In 1962, she was barred from entering Egypt to complete *Cleopatra*; its government announced that "that Miss Taylor will not be allowed to come to Egypt because she has adopted the Jewish faith and 'supports Israeli causes.'" In 1974, Taylor and Richard Burton considered marrying in Israel, but could not because Burton was not Jewish.

Personal Life

Religion and Identity

In 1959, at age 27, after nine months of study, Taylor converted from Christian Science to Judaism, taking the Hebrew name Elisheba Rachel. She stated that her conversion was something she had long considered and was not related to her marriages. After Mike Todd's death, Taylor said that she "felt a desperate need for a formalized religion," and explained that neither Catholicism nor Christian Science were able to address many of the "questions she had about life and death." Biographer Randy Taraborrelli notes that after studying the philosophy of Judaism for nine months, "she felt an immediate connection to the faith."

Marriages and Romances

Taylor was married eight times to seven husbands. When asked why she married so often, she replied, "I don't know, honey. It sure beats the hell out of me," but also said that, "I was taught by my parents that if you fall in love, if you want to have a love affair, you get married. I guess I'm very old-fashioned.

Friendship with Michael Jackson

Taylor and Michael Jackson developed a close friendship. On October 6, 1991, Taylor married construction worker Larry Fortensky at Jackson's Neverland Ranch. In 1997, Jackson presented Taylor with the exclusively written-for-her epic song "Elizabeth, I Love You", performed on the day of her 65th birthday celebration. In 2005, Taylor was a vocal supporter of Jackson in his trial in California on charges of sexually abusing a child. He was eventually cleared of all charges. She encouraged Jackson to wear a Kabbalah red string as protection from the "evil-eye" during the trial. Taylor attended Michael Jackson's private funeral on September 3, 2009.

Illnesses and Death

Taylor struggled with health problems much of her life; starting with her divorce from Hilton, Taylor experienced serious medical issues whenever she faced problems in her personal life. Taylor was hospitalized more than 70 times and had at least 20 major operations. Many times newspaper headlines erroneously announced that Taylor was close to death; she herself only claimed to have almost died on four occasions. At 5'4", Taylor constantly gained and lost significant amounts of weight, reaching both 119 pounds and 180 pounds in the 1980s. She smoked cigarettes into her mid-fifties, and feared she had lung cancer in October 1975 after an X-ray showed spots on her lungs, but was later found not to have the disease. Taylor broke her back five times, had both her hips replaced, had a hysterectomy, suffered from dysentery and phlebitis, punctured her esophagus, survived a benign brain tumor operation in 1997 and skin cancer, and faced life-threatening bouts with pneumonia twice, one in 1961 requiring an emergency tracheotomy. In 1983 she admitted to having been addicted to sleeping pills and painkillers for 35 years. Taylor was treated for alcoholism and prescription drug addiction at the Betty Ford Clinic for seven weeks from December 1983 to January 1984, and again from the autumn of 1988 until early 1989.

On May 30, 2006, Taylor appeared on *Larry King Live* to refute the claims that she had been ill, and denied the allegations that she was suffering from Alzheimer's disease and was close to death. Near the end of her life, however, she was reclusive and sometimes failed to make scheduled appearances due to illness or other personal reasons. She used a wheelchair and when asked about it stated that she had osteoporosis and was born with scoliosis. The mutation that gave Taylor her striking double eyelashes may also have contributed to her history of heart trouble. In 2009 she underwent cardiac surgery to replace a leaky valve. In February 2011, new symptoms related to heart failure caused her to be admitted into Cedars-Sinai Medical Center in Los Angeles for treatment, where she remained until her death at age 79 on March 23, 2011, surrounded by her four children.

Awards and Honors

Taylor won two Academy Awards for Best Actress for her performance in *BUtterfield 8* in 1960, and for *Who's Afraid of Virginia Woolf?* in 1966. Additionally, she received the Jean Herscholt Humanitarian Academy Award in 1992 for her work fighting AIDS. In 1997, Taylor was honored by the Screen Actors Guild (SAG) with the Life Achievement Award.

As we all know, ours was one of the first industries to be directly and dramatically affected by the AIDS epidemic. And it's heartening to me that this community has risen to the challenge. And the foundation of the Screen Actors Guild, of which I'm so proud to be a member, is no exception having made a very generous donation to the Elizabeth Taylor AIDS Foundation. Thank you all for honoring me tonight. Taylor received the French Legion of Honour in 1987, and in 2000 was named a Dame Commander of the Order of the British Empire. In 2001, she received a Presidential Citizens Medal for her humanitarian work, most notably for helping to raise more than $200 million for AIDS research and bringing international attention and resources to addressing the epidemic. Taylor was inducted into the California Hall of Fame in 2007.

73

Virginia Woolf

Adeline Virginia Woolf (25 January 1882 – 28 March 1941) was an English author, essayist, publisher, and writer of short stories, regarded as one of the foremost modernist literary figures of the twentieth century. During the interwar period, Woolf was a significant figure in London literary society and a member of the Bloomsbury Group. Her most famous works include the novels *Mrs Dalloway* (1925), *To the Lighthouse* (1927) and *Orlando* (1928), and the book-length essay *A Room of One's Own* (1929), with its famous dictum, "A woman must have money and a room of her own if she is to write fiction."

Early Life

Virginia Woolf was born Adeline Virginia Stephen in London in 1882. Her mother, a renowned beauty, Julia Prinsep Stephen (born Jackson) (1846–1895), was born in India to Dr. John and Maria Pattle Jackson and later moved to England with her mother, where she served as a model for Pre-Raphaelite painters such as Edward Burne-Jones. Her father, Sir Leslie Stephen, was a notable historian, author, critic and mountaineer. He was the editor of the *Dictionary of National Biography*, a work which would influence Woolf's later experimental biographies. Woolf was educated by her parents in their literate and well-connected household at 22 Hyde Park Gate, Kensington. Her parents had each been married previously and been widowed, and, consequently, the household contained the children of three marriages. Julia had three children from her first husband, Herbert Duckworth: George Duckworth, Stella Duckworth, and Gerald Duckworth. Leslie first married Harriet Marian (Minny) Thackeray (1840–1875), and they had one daughter: Laura Makepeace Stephen, who was declared mentally disabled and lived with the family until she was institutionalised in 1891. Leslie and Julia had four children together: Vanessa Stephen (1879), Thoby Stephen (1880), Virginia (1882), and Adrian Stephen (1883).

Sir Leslie Stephen's eminence as an editor, critic, and biographer, and his connection to William Thackeray (the father of his first wife), meant that his children were raised in an environment filled with the influences of Victorian literary society. Henry James, George Henry Lewes, Julia Margaret Cameron (an aunt of Julia Stephen), and James Russell Lowell, who was made Virginia's honorary godfather, were among the visitors to the house. Julia Stephen was equally well connected. Descended from an attendant of Marie Antoinette, she came from a family of renowned beauties who left their mark on Victorian society as models for Pre-Raphaelite artists and early photographers. Supplementing these influences was the immense library at the Stephens' house, from which Virginia and Vanessa were taught the classics and English literature. Unlike the girls, Adrian and Thoby were formally educated and sent to Cambridge, a difference which Virginia would regret. The sisters did, however, benefit indirectly from their brothers' Cambridge contacts, as the boys brought their new intellectual friends home to the Stephens' drawing room.

Bloomsbury

After the death of their father and Virginia's second nervous breakdown, Vanessa and Adrian sold 22 Hyde Park Gate and bought a house at 46 Gordon Square in Bloomsbury. Woolf came to know Lytton Strachey, Clive Bell, Rupert Brooke, Saxon Sydney-Turner, Duncan Grant, Leonard Woolf and Roger Fry, who together formed the nucleus of the intellectual circle of writers and artists known as the Bloomsbury Group. Several members of the group attained notoriety in 1910 with the Dreadnought hoax, which Virginia participated in disguised as a male Abyssinian royal. Her complete 1940 talk on the Hoax was discovered and is published in the memoirs collected in the expanded edition of *The Platform of Time* (2008). In 1907 Vanessa married Clive Bell, and the couple's interest in avant garde art would have an important influence on Woolf's development as an author. Virginia Stephen married writer Leonard Woolf in 1912. Despite his low material status (Woolf referring to Leonard during their engagement as a "penniless Jew") the couple shared a close bond. Indeed, in 1937, Woolf wrote in her diary: "Love-making – after 25 years can't bear to be separate ... you see it is enormous pleasure being wanted: a wife. And our marriage so complete." The two also collaborated professionally, in 1917 founding the Hogarth Press, which subsequently published Virginia's novels along with works by T.S. Eliot, Laurens van der Post, and others. The Press also commissioned works by contemporary artists, including Dora Carrington and Vanessa Bell.

Work

Woolf began writing professionally in 1900, initially for the *Times Literary Supplement* with a journalistic piece about Haworth, home of the Brontë family. Her first novel, *The Voyage Out*, was published in 1915 by her half-brother's imprint, Gerald Duckworth and Company Ltd. This novel was originally entitled *Melymbrosia*, but Woolf repeatedly changed the draft. An earlier version of *The Voyage Out* has been reconstructed by Woolf scholar Louise DeSalvo and is now available to the public under the intended title. DeSalvo argues that many of the changes Woolf made in the text were in response to changes in her own life.

Death

After completing the manuscript of her last (posthumously published) novel, *Between the Acts*, Woolf fell into a depression similar to that which she had earlier experienced. The onset of World War II, the destruction of her London home during the Blitz, and the cool reception given to her biography of her late friend Roger Fry all worsened her condition until she was unable to work. On 28 March 1941, Woolf put on her overcoat, filled its pockets with stones, and walked into the River Ouse near her home and drowned herself. Woolf's body was not found until 18 April 1941. Her husband buried her cremated remains under an elm in the garden of Monk's House, their home in Rodmell, Sussex.

Modern Scholarship and Interpretations

Recently, studies of Virginia Woolf have focused on feminist and lesbian themes in her work, such as in the 1997 collection of critical essays, *Virginia Woolf: Lesbian Readings*, edited by Eileen Barrett and Patricia Cramer. Controversially, Louise A. DeSalvo reads most of Woolf's life and career through the lens of the incestuous sexual abuse Woolf suffered as a young woman in her 1989 book *Virginia Woolf: The Impact of Childhood Sexual Abuse on her Life and Work*. Woolf's fiction is also studied for its insight into shell shock, war, class and modern British society. Her best-known nonfiction works, *A Room of One's Own* (1929) and *Three Guineas* (1938), examine the difficulties female writers and intellectuals face because men hold disproportionate legal and economic power and the future of women in education and society. Irene Coates's book *Who's Afraid of Leonard Woolf: A Case for the Sanity of Virginia Woolf* holds that Leonard Woolf's treatment of his wife encouraged her ill health and ultimately was responsible for her death. This is not accepted by Leonard's family but is extensively researched and fills in some of the gaps in the traditional account of Virginia Woolf's life. Victoria Glendinning's book *Leonard Woolf: A Biography*, which is even more extensively researched and supported by contemporaneous writings, argues that Leonard Woolf was not only supportive of his wife but enabled her to live as long as she did by providing her with the life and atmosphere she needed to live and write. Virginia's own diaries support this view of the Woolfs' marriage. Though at least one biography of Virginia Woolf appeared in her lifetime, the first authoritative study of her life was published in 1972 by her nephew Quentin Bell. In 1992, Thomas Caramagno published the book *The Flight of the Mind: Virginia Woolf's Art and Manic-Depressive Illness.*"

74

Maya Angelou

Maya Angelou (born April 4, 1928) is an American author and poet who has been called "America's most visible black female autobiographer" by scholar Joanne M. Braxton. She is best known for her series of six autobiographical volumes, which focus on her childhood and early adult experiences. The first and most highly acclaimed, *I Know Why the Caged Bird Sings* (1969), tells of her first seventeen years. It brought her international recognition, and was nominated for a National Book Award. She has been awarded over 30 honorary degrees and was nominated for a Pulitzer Prize for her 1971 volume of poetry, *Just Give Me a Cool Drink of Water 'Fore I Diiie.* Angelou was a member of the Harlem Writers Guild in the late 1950s, was active in the Civil Rights movement, and served as Northern Coordinator of Dr. Martin Luther King, Jr.'s Southern Christian Leadership Conference. Since 1991, she has taught at Wake Forest University in Winston-Salem, North Carolina where she holds the first lifetime Reynolds Professorship of American Studies. Since the 1990s she has made around eighty appearances a year on the lecture circuit. In 1993, Angelou recited her poem "On the Pulse of Morning" at President Bill Clinton's inauguration, the first poet to make an inaugural recitation since Robert Frost at John F. Kennedy's inauguration in 1961. In 1995, she was recognized for having the longest-running record (two years) on *The New York Times* Paperback Nonfiction Bestseller List.

With the publication of *I Know Why the Caged Bird Sings*, Angelou was heralded as a new kind of memoirist, one of the first African American women who was able to publicly discuss her personal life. She is highly respected as a spokesperson for Black people and women. Angelou's work is often characterized as autobiographical fiction. She has, however, made a deliberate attempt to challenge the common structure of the autobiography by critiquing, changing, and expanding the genre. Her books, centered on themes such as identity, family, and racism, are often used as set texts in schools and universities internationally. Some of her more controversial work has been challenged or banned in US schools and libraries.

Early Years

Marguerite Johnson was born in St. Louis, Missouri, on April 4, 1928. Her father, Bailey Johnson, was a doorman and navy dietitian. Her mother, Vivian (Baxter) Johnson, was a real estate agent, trained surgical nurse, and later a merchant marine. Angelou's older brother, Bailey Jr., nicknamed Marguerite "Maya", shortened from "my-a-sister". The details of Angelou's life described in her six autobiographies and in numerous interviews, speeches, and articles tend to be inconsistent. Her biographer, Mary Jane Lupton, has explained that when Angelou has spoken about her life, she has done so eloquently but informally and "with no time chart in front of her". Evidence suggests that Angelou's family is descended from the Mende people of West Africa. A 2008 PBS documentary found that her maternal great-grandmother, Mary Lee, had been emancipated after the Civil War. The documentary suggested that Lee became pregnant by her former white owner, John Savin, who forced Lee to sign a false statement accusing another man of being the father of her child. After indicting Savin for forcing Lee to commit perjury, and despite discovering that Savin was the father, a grand jury found him not guilty. Lee was sent to the Clinton County poorhouse in (Missouri) with her daughter, Marguerite Baxter, who became Angelou's grandmother. Angelou described Lee as "that poor little Black girl, physically and mentally bruised."

The first 17 years of Angelou's life are documented in her first autobiography, *I Know Why the Caged Bird Sings*. When Angelou was three, and her brother four, their parents' "calamitous marriage" ended. Their father sent them to Stamps, Arkansas alone by train to live with her paternal grandmother, Annie Henderson. Henderson prospered financially during the Great Depression and World War II because the general store she owned sold needed basic commodities and because "she made wise and honest investments". Four years later, the children's father "came to Stamps without warning" and returned them to their mother's care in St. Louis. At age eight, while living with her mother, Angelou was sexually abused and raped by her mother's boyfriend, Mr. Freeman. She confessed it to her brother, who told the rest of their family. Freeman was found guilty, but was jailed for one day. Four days after his release, he was found kicked to death, probably by Angelou's uncles. Angelou became mute, believing, as she has stated, "I thought, my voice killed him; I killed that man, because I told his name. And then I thought I would never speak again, because my voice would kill anyone..." She remained mute for nearly five years. Shortly after Freeman's murder, Angelou and her brother were sent back to their grandmother once again.

Adulthood and Early Career

Angelou has been married three times or more (something she has never clarified, "for fear of sounding frivolous"). In her third autobiography, *Singin' and Swingin' and Gettin' Merry Like Christmas*, Angelou describes her three-year marriage to Greek sailor Tosh Angelos in 1949. Up to that point she went by the name of "Marguerite Johnson", or "Rita", but changed her professional name to "Maya Angelou." Her managers at San Francisco nightclub The Purple Onion strongly suggested that she adopt a more theatrical name that captured the feel of her Calypso dance performances. In 1952, she won a scholarship and trained in African dance with dancer Pearl Primus of Trinidad. Later Angelou studied modern dance with Martha Graham.

Later Career

In 1973, Angelou married Paul du Feu, a British-born carpenter and remodeler, and moved to Sonoma, California with him. The years to follow were some of Angelou's most productive as a writer and poet. She worked as a composer, writing for singer Roberta Flack and composing movie scores. She wrote articles, short stories, TV scripts, autobiographies and poetry, produced plays, and spoke on the university lecture circuit. In 1977 Angelou appeared in a supporting role in the television mini-series *Roots*. Her screenplay, *Georgia, Georgia*, was the first original script by a Black woman to be produced. In the late '70s, Angelou met Oprah Winfrey when Winfrey was a TV anchor in Baltimore, Maryland; Angelou would later become Winfrey's close friend and mentor. Angelou divorced de Feu and returned to the southern United States in 1981, where she accepted the first lifetime Reynolds Professorship of American Studies at Wake Forest University in Winston-Salem, North Carolina. In 1993, she recited her poem *On the Pulse of Morning* at the inauguration of President Bill Clinton, becoming the first poet to make an inaugural recitation since Robert Frost at John F. Kennedy's inauguration in 1961. Since the 1990s, Angelou has actively participated in the lecture circuit. Angelou campaigned for the Democratic Party in the 2008 presidential primaries, giving her public support to Senator Hillary Clinton. When Clinton's campaign ended, Angelou put her support behind Senator Barack Obama, who won the election and became the first African American president of the United States. She stated, "We are growing up beyond the idiocies of racism and sexism". In 2009, Angelou campaigned for the same-sex marriage bill in New York state. Angelou was the first African American woman to direct a major motion picture, *Down in the Delta*, in 1998, at the age of seventy.

Angelou's Work

Although Angelou wrote her first autobiography, *I Know Why the Caged Bird Sings*, without the intention of writing a series, she went on to write five additional volumes. They are distinct in style and narration. The volumes "stretch over time and place", from Arkansas to Africa and back to the US. They take place from the beginnings of World War II to King's assassination. As author Lyman B. Hagen states, Angelou has "opened her life to public scrutiny through her works". Like *Caged Bird*, the events in these books are episodic and crafted like a series of short stories, but do not follow a strict chronology. Later books in the series include *Gather Together in My Name* (1974), *Singin' and Swingin' and Gettin' Merry Like Christmas* (1976), *The Heart of a Woman* (1981), *All*

God's Children Need Traveling Shoes (1986), and *A Song Flung Up to Heaven* (2002). Angelou's book of essays, *Wouldn't Take Nothing For My Journey Now* (1993), contains materials that are autobiographical in content. Critics have tended to judge Angelou's subsequent autobiographies "in light of the first", with *Caged Bird* receiving the highest praise. Angelou has used the same editor throughout her writing career, Robert Loomis, an executive editor at Random House, who has been called "one of publishing's hall of fame editors." Angelou has said regarding Loomis: "We have a relationship that's kind of famous among publishers".

Themes in Angelou's Autobiographies

As feminist scholar Maria Lauret has indicated, Angelou and other female writers in the late 1960s and early 1970s used the autobiography to reimagine ways of writing about women's lives and identities in a male-dominated society. Lauret has made a connection between Angelou's autobiographies, which Lauret called "fictions of subjectivity" and "feminist first-person narratives", and fictional first-person narratives (such as *The Women's Room* by Marilyn French and *The Golden Notebook* by Doris Lessing) written during the same period. Both genres employ the narrator as protagonist and "rely upon the illusion of presence in their mode of signification". Lauret has also stated that "the formation of female cultural identity" has been woven into Angelou's narratives. Angelou has presented herself as a role model for African American women by reconstructing the Black woman's image throughout her autobiographies, and has used her many roles, incarnations, and identities to "signify multiple layers of oppression and personal history". Lauret has viewed Angelou's themes of the individual's strength and ability to overcome throughout Angelou's autobiographies as well.

One of the most important themes in Angelou's autobiographies are "kinship concerns", from the character-defining experience of her parents' abandonment to her relationships with her son, husbands, and lovers throughout all of her books. African American literature scholar Dolly McPherson has insisted that Angelou's concept of family throughout her books must be understood in the light of the way in which she and her older brother were displaced by their parents at the beginning of *Caged Bird*. Motherhood is a "prevailing theme" in all of Angelou's autobiographies, specifically her experiences as a single mother, a daughter, and a granddaughter. Lupton believes that Angelou's plot construction and character development were influenced by this mother/child motif found in the work of Harlem Renaissance poet Jessie Fauset. Scholar Sondra O'Neale agrees, and insists that Angelou's autobiographies present black women differently than literature had portrayed them up to that time. O'Neale goes on to state that "no Black woman in the world of Angelou's books are losers", and that Angelou herself is the third generation of "brilliantly resourceful females" who overcame the obstacles of racism and oppression. Lupton has stated that the one unifying theme that connects all of Angelou's autobiographies is what she has called "the mother-child pattern". Angelou describes throughout her books her connection of mother and child—with herself and her son Guy, with herself and her own mother, and with herself and her grandmother.

Awards and Honors

Angelou is one of the most honored writers of her generation. She has been honored by universities, literary organizations, government agencies, and special interest groups. Her honors include a National Book Award nomination for *I Know Why the Caged Bird Sings,* a Pulitzer Prize nomination for her book of poetry, *Just Give Me a Cool Drink of Water 'fore I Diiie,* a Tony Award nomination for her role in the 1973 play *Look Away*, and three Grammys for her spoken word albums. In 1995, Angelou's publishing company, Bantam Books, recognized her for having the longest-running record (two years) on *The New York Times* Paperback Nonfiction Bestseller List. In 1998, she was inducted into the National Women's Hall of Fame. She has served on two presidential committees, and was awarded the National Medal of Arts in 2000, the Lincoln Medal in 2008, and the Presidential Medal of Freedom in 2011. Musician Ben Harper has honored Angelou with his song "I'll Rise", which includes words from her poem, "Still I Rise." She has been awarded over thirty honorary degrees.

75

Anita Roddick

Dame Anita Roddick, DBE (23 October 1942 – 10 September 2007) was a British businesswoman, human rights activist and environmental campaigner, best known as the founder of The Body Shop, a cosmetics company producing and retailing beauty products that shaped ethical consumerism. The company was one of the first to prohibit the use of ingredients tested on animals and one of the first to promote fair trade with third world countries. Roddick was involved in activism and campaigning for environmental and social issues including involvement with Greenpeace and The Big Issue. In 1990, Roddick founded Children On The Edge, a charitable organization which helps disadvantaged children in Eastern Europe and Asia. In 2003, Queen Elizabeth II appointed Roddick a Dame Commander of the Order of the British Empire. Anita Roddick was born in a bomb shelter in Littlehampton, Sussex the daughter of Italian immigrants who had recently arrived in England from Naples. After school, Anita trained as a teacher before working for the United Nations. This experience enabled her to travel extensively and meet many many people. In 1979, she opened her first shop in Brighton known as 'The Shop' located between two funeral parlours, Anita Roddick said that many business decisions came about through circumstances and good luck rather than any prepared plan. The Body Shops simple ethical message and promotion of ethical and environmental friendly consumerism helped grow a very strong brand image. By 1991, the Body Shop had over 700 shops worldwide. It is considered to be the second most respected brand name in the UK. In 1990, after visiting a Romanian orphanage, she established a charity - Children On The Edge. She actively campaigned for this and other charities. In 2006, she sold the Body Shop to L'Oreal for £652million. In 2007, Anita disclosed she was quiet ill with Hepatitis C shortly after she died from a brain haemorrhage. She left her estate to charities.

76

Dame Barbara Cartland

> To sleep around is absolutely wrong for a woman; it's degrading and it completely ruins her personality. Sooner or later it will destroy all that is feminine and beautiful and idealistic in her.
>
> —*Barbara Cartland*

Dame Barbara Cartland was a prolific and popular write of Romantic novels. Using a tried and tested formula, she tapped into a populist demand for romantic novels, usually with women ending up happily married to some chivalrous man. Outside of writing she was a vivacious and outgoing personality, rarely short of an opinion on any subject. She was also noted for her dress sense, which was nothing if not colourful. She was born in Edgbaston, Birmingham in 1901. Her father died during the First World War, leaving the family in difficult circumstances. However, with her natural charm, beauty and energy, she became a feature on the 1920s social circle. She attracted many proposals of marriage (said to be 49) before finally accepting. Ironically, her first marriage was not a success, ending in divorce 6 years later. However, her second marriage to Alexander McCorquodale was successful and she had two children through this marriage. In the 1920s, Barbara Cartland became friendly with Lord Beaverbrook (though she refused to be his mistress). This led to a position at the Daily Mail where she gained a column to write on issues of the day. In 1923, she wrote her first novel Jigsaw. The success of this book led greater demand from publishers. She had found a formula for writing and could dictate a new novel in one fortnight at the peak of her writing efficiency. By the end of her life she had written an estimated 723 books.

Apart from a few, the majority of books were based on an old fashioned romantic vision of love. Her women were models of virtue and her men had the old fashioned virtues of chivalry. Her writing hardly endeared herself to feminists or literary critics. A woman asking "Am I good? Am I satisfied?" is extremely selfish. The less women fuss about themselves, the less they talk to other women, the more they try to please their husbands, the happier the marriage is going to be. But, the popularity of the books could not be denied. She was a colourful personality and during the Royal Wedding of Charles and Diana in 1981, it was feared she would upstage the young couple, through her relationship with Lady Diana. However, she avoided attending the wedding to avoid taking the spotlight. She was also a defender of old fashioned beliefs; she believed that women should remain chaste until marriage.

Though paradoxically, she believed men should have experience (when asked how this was possible she replied men could visit brothels to get the necessary experience.) Yet, life wasn't all idealised romantic fantasies. She suffered great personal loss in both first world Wars. After helping her brother Ronald Cartland to become an MP, he became the first MP to die during the conflict at the battle of Dunkirk. Later her elder brother died. After this she temporarily retired to Canada, but felt she should return to England to take part in the war effort. She joined the women's voluntary service and worked for several years for St Johns Ambulance. Dame Barbara Cartland became a national treasure. To som, maybe a caricature of previous times - even a figure of fun. But, in becoming one of the top 10 best selling authors of all time, she made sure she was having the last laugh.

77

Beatrix Potter

Early Life

Born to a comfortable middle class family, Beatrix Potter became one of the best selling children's author of all time. Devoted to her beloved lake district, she also donated over 4,000 acres of Lake farmland to the National Trust. She spent much of her early life in her own company; she rarely saw her brother Ewan, who was sent to boarding school. Having little social contact with children of her own age, Beatrix began to be drawn into her own world of creating her own stories, based on animals. Beatrix was a naturally gifted artist, and with some art lessons she also learnt the technical side of drawing. During her childhood, and especially in the Lake District, she looked after many animals, such as; rabbits, frogs, and even bats. She drew these animals throughout her childhood, gradually improving the standard of her drawings. Beatrix was also interested in natural history; she would spend many hours drawing wildlife such as fungi and flowers. At one time she had aspirations to develop this scientific interest. An uncle tried to help her become a student at the Royal Botanic Gardens at Kew, but she was rejected because of her gender. Nevertheless, she became respected for her contribution to mycology -the study of fungi

In her early 20s, Beatrix's parents tried to arrange a suitable partner for Beatrix to marry. Many suitable suitors were found, however for each prospective marriage partnership Beatrix turned them down. She was a fiercely independent woman, and she disliked the idea of being tied down to an uneventful domestic life of staying at home and bringing up children. Thus, unusually for the late Victorian time period, Beatrix remained single and stayed at home.

Publication of Peter Rabbit Books

In her 20s that she sought to try and get her children's book and drawings published. Her initial attempts proved unsuccessful, but she persevered and eventually it was taken on by Frederick Warne & Company. The book was published in 1902, when Beatrix was 36. The publishers did not have much hope it would sell many copies; they actually gave the project to their youngest brother, Norman, as a kind of test for his first project. However, Norman proved to be a good choice. He warmed to both the book and Beatrix. He was determined to make a success of the book and developed a good working relationship with Beatrix as they pored over the individual details of the book. It was Norman who insisted that each drawing of Peter Rabbit would be in colour. Beatrix insisted that the book remain small, so that it would be easy for children to hold. By the end of the year, 28,000 copies were in print.

Relationship with Norman Warne

The relationship between Norman and Beatrix blossomed, and eventually they became engaged. However, Beatrix's parents disapproved. They felt it wrong for Beatrix to marry a tradesman. However, they eventually relented, but insisted Beatrix live apart for 6 months; giving her time to change her mind. Tragically, before the wedding could take place, Norman passed away, dyeing of pernicious anaemia. Beatrix was devastated, she wrote a letter to his sister, Millie, saying; "He did not live long, but he fulfilled a useful happy life. I must try to make a fresh beginning next year." After his death, Beatrix moved to the Lakeland. In 1905, she bought Hill Top farm, in Sawry, Cumbria. She lived here for the remainder of her life. Due to failing eyesight, Beatrix later stopped writing her children books; instead, she devoted her time to the breading of sheep and helping the conservation of Lakeland farms.

Beatrix Potter - Conservation in Lake District

Due to proceeds from her very successful books and later her inheritance, Beatrix was able to buy many working farms.

On her death she left over 4,000 acres to the National Trust. It is one of the biggest legacy's ever made. Potter wrote 23 Books. Some of her best know titles include:

- The Tale of Peter Rabbit (1902)
- The Tale of Squirrel Nutkin (1903)
- The Tailor of Gloucester (1903)
- The Tale of Benjamin Bunny (1904)
- The Tale of Two Bad Mice (1904)
- The Tale of Mrs. Tiggy-Winkle (1905)
- The Tale of the Pie and the Patty-Pan (1905)
- The Tale of Mr. Jeremy Fisher (1906)
- The Story of A Fierce Bad Rabbit (1906)

78

Shakira

Born in Columbia in 1997, Shakira Isabel Mebarak Ripoll, displayed a prodigious musical talent from an early age. Her father encouraged her to write and create song lyrics. Even by the age of 7 she showed remarkable talent in both singing and dancing. It was in her grandmother's Lebanese restaurant where she learnt and developed dancing routines. Although born in Columbia her father was Lebanese. As a child her voice was very powerful and some criticised it for being 'too powerful' But, Shakira caught the eye of music producers and she started recording her own albums. In 1995, she released Pies Descalzos', which gained her much popularity in Latin America and Spain Her 1998 album Dónde Están Los Ladrones? was very influential and moved her from a local star to a global fan base and international appeal. By the early 2000s, she had broken into the market for English speaking music. She learnt English and wrote and recorded her own songs, producing albums in both Spanish and English. She has set numerous records for her albums, downloads and has become easily the best selling Colombian artist of all time.

As well as music, she is involved in humanitarian projects, especially in her native Columbia which is struggling to recover from long periods of civil war. Shakira has offered money for her educational charities (Bare Feet Foundation, named after her first hit album), but, equally important she has offered her public presence which often galvanises politicians into promising to do more. When asked why she spends considerable time on humanitarian projects she replied that she felt it was important after having grown up in a developing country and being aware of the huge gap between rich and poor. She is one of the top 5 best selling artists of the 2000s. Shakira also performed at Barack Obama's inaugural ceremony in 2009.

79

Anna Pavlova

Anna Pavlova (January 31, 1881 – January 23, 1931) was a Russian ballerina of the late 19th and the early 20th century. She is widely regarded as one of the finest classical ballet dancers in history and was most noted as a principal artist of the Imperial Russian Ballet and the Ballets Russes of Sergei Diaghilev. Pavlova is most recognised for the creation of the role *The Dying Swan* and, with her own company, would become the first ballerina to tour ballet around the world.

Early Life

Pavlova was born prematurely on January 31, 1881, in Ligovo, a suburb (now neighborhood) of Saint Petersburg, then the capital of the Russian Empire. Her mother was a laundress named Lyubov Feodorovna. The identity of her father has been open to debate. She later claimed her father (who was of possible Jewish origin) had died when she was two years old. Some sources, including *The Saint Petersburg Gazette*, have claimed that her illegitimate father was the banker Lazar Polyakov. Her mother's second husband, Matvey Pavlov, is believed to have adopted her at the age of three, by which she acquired her last name. Pavlova's passion for the art of ballet was ignited when her mother took her to a performance of Marius Petipa's original production of *The Sleeping Beauty* at the Imperial Maryinsky Theater. The lavish spectacle made an impression on the young Pavlova, and at the age of eight her mother took her to audition for the renowned Imperial Ballet School. She was not chosen due to her age and for what was considered to be a "sickly" appearance, but she was finally accepted at the age of 10 in 1891. She appeared for the first time on stage in Marius Petipa's *Un conte de fées* (*A Fairy Tale*), which the ballet master staged for the students of the school.

The young Pavlova's years of training were difficult, as classical ballet did not come easily to her. Her severely arched feet, thin ankles, and long limbs clashed with the small and compact body in favor for the ballerina at the time. Her fellow students taunted her with such nicknames as *The broom* and *La petite sauvage* (The little savage). Undeterred, Pavlova trained to improve her technique. She took extra lessons from the noted teachers of the day — Christian Johansson, Pavel Gerdt, Nikolai Legat and more especially from Enrico Cecchetti, considered the greatest ballet virtuoso of the time and founder of the Cecchetti method, a very influential ballet technique used up to this day. In 1898 she entered the *classe de perfection* of Ekaterina Vazem, former *Prima ballerina* of the Saint Petersburg Imperial Theatres. During her final year at the Imperial Ballet School, she performed many roles with the principal company. She graduated in 1899 at age 18, being allowed to enter the Imperial Ballet a rank ahead of *corps de ballet* as a *coryphée*. She made her official début at the Mariinsky Theatre in Pavel Gerdt's *Les Dryades prétendues* (*The False Dryads*). Her performance drew praise from the critics, particularly the great critic and historian Nikolai Bezobrazov.

Personal Life

Victor Dandré, her manager and companion, may have been her husband (she deliberately clouded this issue).

Death

While touring in The Hague, Netherlands, Pavlova was told that she had pneumonia and needed an operation. She was also told that she would never be able to dance again if she had this operation. She refused to have the operation saying "If I can't dance then I'd rather be dead." Three weeks later she died of pleurisy, eight days short of her 50th birthday. She was holding her costume from *The Dying Swan* when she spoke her last words, "Play the last measure very softly." Her death came in the Hotel Des Indes in The Hague, which displays a wall plaque and has a cigar lounge named the

Anna Pavlova Library in her memory. In accordance with old ballet tradition, on the day she was to have next performed, the show went on as scheduled, with a single spotlight circling an empty stage where she would have been. Memorial services were held in the Russian Orthodox Church in London. Anna Pavlova was cremated, and her ashes placed in a columbarium at Golders Green Crematorium, where her urn was subsequently adorned with her ballet shoes (which since then have been stolen).

Legacy

Pavlova inspired the choreographer Frederick Ashton when as a boy of 13 he saw her dance in in the Municipal Theater in Lima, Peru. The Pavlova dessert is believed to have been created in honour of the dancer either during or after one of her tours to Australia and New Zealand in the 1920s. The nationality of its creator has been a source of argument between the two nations for many years. The Jarabe Tapatío, known in English as the 'Mexican Hat Dance', gained popularity outside of Mexico when Pavlova created a staged version in pointe shoes, for which she was showered with hats by her adoring Mexican audiences. Afterward, in 1924, the Jarabe Tapatío was proclaimed Mexico's national dance. She once said that the country that would produce the best ballerina in history would be the United States because of all the different cultures that came together there. Anna Pavlova was able to complete 37 twirls while on top of a moving elephant while on a tour in China. Pavlova's life was depicted in the 1983 film *Anna Pavlova*. When the Victoria Palace Theatre in London, England, opened in 1911, a gilded statue of Pavlova had been installed above the cupola of the theatre. This was taken down for its safety during World War II and was lost. In 2006, a replica of the original statue was restored in its place.

80

Rani Lakshmibai

Lakshmi Bai, the Rani of Jhansi (c.19 November 1835 – 17 June 1858), known as Jhansi Ki Rani, or the queen of Jhansi, was one of the leading figures of the Indian Rebellion of 1857, and a symbol of resistance to British rule in India. She was the queen of the Maratha-ruled princely state of Jhansi, situated in the northern part of India.

Early life

Originally named Manikarnika and nicknamed Manu, she was born on 19 November 1835 at Kashi (Varanasi) to a Maharashtrian Marathi Karhade Brahmin family, the daughter of Moropant Tambe and Bhagirathibai Tambe. She lost her mother at the age of four, and was educated at home. Her father, Moropant Tambe, worked at the court of Peshwa at Bithur, who brought her up like his own daughter, and called her "Chhabili" because of her light-heartedness. Because of her father's influence at court, Rani Lakshmi Bai had more independence than most women, who were normally restricted to the *zenana.* She studied self defence, horsemanship, archery, and even formed her own army out of her female friends at court. She was married to Raja Gangadhar Rao Newalkar, the Maharaja of Jhansi in 1842, and became the queen of Jhansi. After their marriage, she was given the name Lakshmi Bai. She gave birth to a son Damodar Rao in 1851. However, the child died when he was about four months old. After the death of their son, the Raja and Rani of Jhansi adopted Anand Rao. Anand Rao was the son of Gangadhar Rao's cousin. However, it is said that the Raja of Jhansi never recovered from his son's death, and he died on 21 November

1853. Because Anand Rao was adopted and not biologically related to the Raja, the East India Company, under Governor-General Lord Dalhousie, had an excuse to apply the Doctrine of Lapse, rejecting Rao's claim to the throne. Dalhousie then annexed Jhansi, saying that the throne had "lapsed" and claimed the right to put Jhansi under his protection. In March 1854, she was given a pension of 60,000 rupees and ordered to leave the palace at the Jhansi fort.

The 1857 Revolution

While this was happening in Jhansi, on May 10, 1857 the Indian Rebellion started in Meerut. This became the starting point for the rebellion against the British. It began after rumours that the new bullet casings for the Lee Enfield rifles were coated with pork and beef fat. British commanders insisted on their use and started to discipline anyone who disobeyed. During this rebellion many British soldiers and officers of the East India Company were killed by the sepoys. Meanwhile, unrest began to spread throughout India and, in May 1857, the Indian Rebellion erupted in numerous pockets across the northern subcontinent. During this chaotic time, the British were forced to focus their attentions elsewhere, and Lakshmi Bai was essentially left to rule Jhansi alone. During this time, she was able to swiftly and efficiently lead her troops against skirmishes breaking out in Jhansi. Through this leadership Lakshmi Bai was able to keep Jhansi relatively calm and peaceful in the midst of the Empire's unrest. For example, she conducted the haldi-kumkum ceremony with great pomp for all the women of Jhansi to provide assurance to her subjects and to convince them that Jhansi was under no threat of an attack. Up to this point, she had been hesitant to rebel against the British, and there is still some controversy over her role in the massacre of the British HEIC officials and their wives and children on the 8th June 1857 at Jokhan Bagh. Her hesitation finally ended when British troops arrived under Sir Hugh Rose and laid siege to Jhansi on 23 March 1858. She rallied her troops around her and fought fiercely against the British. An army of 20,000, headed by the rebel leader Tatya Tope, was sent to relieve Jhansi and to take Lakshmi Bai to freedom. However, the British, though numbering only 1,540 in the field so as not to break the siege, were better trained and disciplined than the raw recruits, and these inexperienced soldiers turned and fled shortly after the British began to attack on 31st March. Lakshmi Bai's forces could not hold out and three days later the British were able to breach the city walls and capture the city. Lakshmi Bai escaped by jumping from the wall at night and fled from her city, surrounded by her guards, many of whom were women from her military. Along with the young Anand Rao, the Rani decamped to Kalpi along with her forces where she joined other rebel forces, including those of Tatya Tope. The Rani and Tatya Tope moved on to Gwalior, where the combined rebel forces defeated the army of the Maharaja of Gwalior after his armies deserted the rebel forces. They then occupied a strategic fort at Gwalior. However, on either the 17th or 18th of June 1858, while battling in full warrior regalia against the 8th King's Royal Irish Hussars in Kotah-ki Serai near the Phool Bagh area of Gwalior, she died. The British captured Gwalior three days later. In the British report of the battle, General Hugh Rose commented that the Rani, "remarkable for her beauty, cleverness and perseverance", had been "the most dangerous of all the rebel leaders". Her father, Moropant Tambey, was captured and hanged a few days after the fall of Jhansi. Her adopted son, Damodar Rao (formerly known as Anand Rao), fled with his mother's aides. Rao was later given a pension by the British Raj and cared for, although he never received his inheritance. Damodar Rao settled down in the city of Indore (Madhya Pradesh). He spent most of his life trying convince the British to restore some of his rights. He and his descendants took on the last name Jhansiwale. He died on May 28, 1906, aged 58.

Legacy

Rani Lakshmi Bai became a national heroine and was seen as the epitome of female bravery in India. When Subhas Chandra Bose's Indian National Army created its first female unit, it was named after her.

Indian poetess Subhadra Kumari Chauhan (1904–1948) wrote a poem titled *Jhansi Ki Rani* in the *Veer Ras* style about her.

81

Sonia Gandhi

Sonia Gandhi (born Edvige Antonia Albina Maino on 9 December 1946) is the President of Indian National Congress, one of the major political parties of India. She is Italian-born daughter-in-law of the late Prime Minister of India, Mrs. Indira Gandhi. After her husband Rajiv Gandhi's assassination in 1991, she was invited by the Indian Congress Party to take over the Congress but Gandhi refused and publicly stayed away from politics amidst constant prodding by the Congress. She finally agreed to join politics in 1997 and in 1998, she was elected as the leader of the Congress. Since then, Gandhi has been the President of the Indian National Congress Party. She has served as the Chairperson of the ruling United Progressive Alliance in the Lok Sabha since 2004. In September 2010, on being re-elected for the fourth time, she became the longest serving president in the 125-year history of the Congress party. Her foreign birth has been a subject of much debate and controversy. Although Sonia is actually the fifth foreign-born person to be leader of the Congress Party, she is the first since independence in 1947.

Early Life

She was born to Stefano and Paola Maino in contrada Màini ("Maini street") in Lusiana, a little village 30 km from Vicenza in Veneto, Italy. She spent her adolescence in Orbassano, a town near Turin, being raised in a traditional Roman Catholic family and attending a Catholic school. Her father, a building contractor, died in 1983. Her mother and two sisters still live around Orbassano. In 1964, she went to study English at the Bell Educational Trust's language school in the city of Cambridge. She met Rajiv Gandhi, who was enrolled in Trinity College at the University of Cambridge in 1965 at a Greek restaurant while working there, as a waitress to make ends meet. Sonia and Rajiv Gandhi married in 1968, following which she moved into the house of her mother-in-law and then Prime Minister, Indira Gandhi. The couple had two children, Rahul Gandhi (born 1970) and Priyanka Vadra (born 1972). Despite belonging to the influential Nehru family, Sonia and Rajiv avoided all involvement in politics. Rajiv worked as an airline pilot while Sonia took care of her family. When Indira was ousted from office in 1977 in the aftermath of the Indian Emergency, the Rajiv family moved abroad for a short time. When Rajiv entered politics in 1982 after the death of his younger brother Sanjay Gandhi in a plane crash on 23 June 1980, Sonia continued to focus on her family and avoided all contact with the public.

Political Career

Wife of the Prime Minister

Sonia Gandhi's involvement with Indian public life began after the assassination of her mother-in-law and her husband's election as Prime Minister. As the Prime Minister's wife she acted as his official hostess and also accompanied him on a number of state visits. In 1984, she actively campaigned against her husband's sister-in-law Maneka Gandhi who was running against Rajiv in Amethi. At the end of Rajiv Gandhi's five years in office, the Bofors Scandal broke out. Ottavio Quattrocchi, an Italian business man believed to be involved, was said to be a friend of Sonia Gandhi, having access to the Prime Minister's official residence. In 1980, her name appeared in the voter's list for New Delhi prior to her becoming an Indian Citizen, when she was still holding Italian Citizenship. It was a violation of Indian Laws. When she did acquire Indian Citizenship in April 1983, the issue cropped up again, as her name appeared on the 1983 voter's list when the deadline for registering had been in January 1983. Senior Congress leader Pranab Mukherjee said that she surrendered her Italian passport to the Italian Embassy on 27 April 1983. Italian nationality law did not permit dual nationality until 1992. So, by acquiring Indian citizenship in 1983, she would automatically have lost Italian citizenship.

Congress President

After the assassination of Rajiv Gandhi and her refusal of becoming Prime Minister, the party settled on the choice of P. V. Narasimha Rao who became leader and subsequently Prime Minister. Over the next few years, however, the Congress fortunes continued to dwindle and it lost the 1996 elections. Several senior leaders such as Madhavrao Sindhia, Rajesh Pilot, Narayan Dutt Tiwari, Arjun Singh, Mamata Banerjee, G. K. Moopanar, P. Chidambaram and Jayanthi Natarajan were in open revolt against incumbent President Sitaram Kesri and quit the party, splitting the Congress into many factions.

Leader of the Opposition

She was elected the Leader of the Opposition of the 13th Lok Sabha in 1999. When the BJP-led NDA formed a government under Atal Bihari Vajpayee, she took the office of the Leader of Opposition. As Leader of Opposition, she called a no-confidence motion against the NDA government led by Vajpayee in 2003. She holds the record of having served as Congress President for 10 years consecutively.

UPA Chairperson

On 23 March 2006, Gandhi announced her resignation from the Lok Sabha and also as chairperson of the National Advisory Council under the office-of-profit controversy and the speculation that the government was planning to bring an ordinance to exempt the post of chairperson of National Advisory Council from the purview of office of profit. She was re-elected from her constituency Rae Bareilly in May 2006 by a margin of over 400,000 votes. As chairperson of the National Advisory Committee and the UPA, she played an important role in making the National Rural Employment Guarantee Scheme and the Right to Information Act into law.

Personal Life

Sonia is the widow of late Rajiv Gandhi, elder son of Indira Gandhi. There has been considerable media speculation for over a decade about their future role in the Congress. After a period of uncertainty, both Rahul and Priyanka became primary members of the Congress party. While Priyanka has so far restricted herself to organizing her mother's election campaigns and taking care of Sonia's constituency, Rahul Gandhi has gone on to take formal charge as General Secretary of the Congress Party. He is also currently head of the Youth Congress.

Honours, Awards and International Recognition

Gandhi was named the third most powerful woman in the world by Forbes Magazine in the year 2004 and was ranked 6th in 2007. In 2010, Gandhi ranked as the ninth most powerful person on the planet by *Forbes Magazine*. She was also named among the *Time 100 most influential* people in the world for the years 2007 and 2008. The British magazine *New Statesman* listed Sonia Gandhi at number 29 in their annual survey of "The World's 50 Most Influential Figures" in the year 2010.

82

Edith Cavell

Edith Louisa Cavell (4 December 1865 - 12 October 1915). Edith Cavell was a nurse, humanitarian and spy. During the First World War, she helped allied servicemen escape German occupied Belgium; she was eventually captured and executed for treason. Her death by firing squad made her internationally known and she became an iconic symbol for the Allied cause. In particular, she is remembered for her courage in facing execution with equanimity. This included her famous last words that '*Patriotism is not enough.*' Edith Cavell was born in Swardeston, near Norwich. Her father was a priest in the Anglican church; this religious faith, she was brought up with, was to provide an important influence on her life. In 1900, she trained to be a nurse at the London hospital. In 1907, she was recruited to be the matron of a new nursing school in Brussels. This was a period of growth in the prestige and importance of nursing; a period which began with Florence Nightingale during the Crimean War. In 1910, Miss Cavell began one of the first nursing journals, *L'infirmiere*, this documented good nursing practises and basic standards. She became a teacher of nurses in different hospitals throughout Belgium and sought to improve standards of nursing. In the Nursing Mirror, Edith Cavell, writes:

> "The probationers wear blue dresses with white aprons and white collars. The contrast which they present to the nuns, in their heavy stiff robes, and to the lay nurses, in their grimy apparel, is the contrast of the unhygienic past with the enlightened present."

Edith Cavell - First World War

In 1914, the First World War broke out. At the time, Miss

Cavell was in England, but she moved back to Belgium to her hospital which was later taken over by the Red Cross. As part of the German Schlieffen plan, the Germans invaded Belgium and from late 1914, Brussels was under a very strict German occupation of military rule. Many British soldiers had been lost behind in the withdrawal of the allied forces and were stuck in Brussels. Miss Cavell decided to aid the British servicemen, hiding them in the hospital and safe houses around Belgium. From these safe houses, some 200 British servicemen were able to escape to neutral Holland. At the same time, she continued to act as nurse and treated wounded soldiers from both the German and allied side. The occupying German army threatened strict punishments for anyone who was found to be 'aiding and abetting the enemy'. Yet, despite the military rule, Miss Cavell continued to help.

> "Nothing but physical impossibility, lack of space and money would make me close my doors to Allied refugees."
>
> —*Edith Cavell*

In mid 1915, nurse Edith Cavell came under suspicion for helping allied servicemen to escape; this was not helped by her outspoken views on her perceived injustice of the occupation. In August 1915, she was arrested and held in St Gilles prison. After her arrest, she did not try to defend herself, but only said in her defense that she felt compelled to help the people in need. After a short trial, the German military tribunal found her guilty of treason and sentenced her to execution. This surprised many observers as it seemed harsh given her honesty and fact she had saved many lives both Allied and German. Brand Whitlock, the US minister to Belgium and the Spanish Minister, 'The Marquis de Villalobar', made representations to the German High Command asking her sentence of death be commuted. In particular, the US minister warned the Germans that this execution of a nurse would damage Germany's already bad reputation and would be seen as an injustice in the eyes of the world. However, the protestations from the Spanish and American embassies were in vain, the German officer in charge - Count Harrach, dismissed the pleas saying *'He would rather see Miss Cavell shot than have harm come to one of the humblest German soldiers, and his only regret was that they had not 'three or four English old women to shoot.'*

Execution of Nurse Edith Cavell

For two weeks prior to her execution, Miss Cavell, was kept in solitary confinement, except for a few brief visits. On the night before her execution, she was visited by the Reverend Stirling Gahan, an Anglican chaplain. He recorded her final conversation.

The following morning she was executed with other Belgians convicted on similar charges. There are conflicting reports of her execution. But, in one report, a German soldier is said to have refused to execute Miss Cavell and was shot by his commanding officer. However, this account is refuted by Pasteur Le Seur who was at the execution. In some misleading allied propaganda, Edith Cavell was reported to have fainted with fear and refused to wear a blindfold, after which she was shot in head by a German officer. This was found to be untrue.

Edith Cavell and War Propaganda

After her execution, the fate of Edith Cavell was widely publicised in the British and American media. It was shown as more evidence of German brutality and injustice. Edith Cavell was portrayed as a heroic, and innocent figure who remained steadfast in her Christian faith and willingness to die for her country. It was hoped her example would encourage more men to enlist. The incident and disgust at her treatment by Germany, played an important role in shaping American public opinion and easing America's entry into the war, later in 1917. Interestingly, during the war the French shot two German nurses helping German forces escape. When asked why they didn't publicise this for its similarities to Edith Cavell's execution, the German High Command replied, 'Why complain? the French had a perfect right to shoot them.' After the war her body was returned to Westminster abbey for a state burial. Her body was later buried in Norwich Cathedral.

83

St. Therese of Lisieux

"Our fulcrum is God: our lever, prayer; prayer which burns with love. With that we can lift the world!"

—St Therese

From an early age it was Therese's ambition and desire to be a saint. She was born into a pious and loving Catholic family. She remembers the idyll of her early childhood, spending time with her parents and 5 sisters in the un spoilt French countryside. However this early childhood idyll was broken by the early death of her Mother (from breast cancer). Aged only 4 years old, she felt the pain of separation and instinctively turned to the Virgin Mary for comfort and reassurance. The next couple of years of St Therese's' life was a period of inner turmoil. She was unhappy at school, where her natural precociousness and piety, made other school children jealous. Eventually her father agreed for Therese to return home and be taught by her elder sister, Celine. She enjoyed being taught at home, however after a while, her eldest sister made a decision to leave to enter the local Carmel Convent at Lisieux. This made Therese feel like she had lost her second mother. Shortly afterwards Therese experienced a painful illness, in which she suffered delusions. The doctors were at a loss as to the cause. For 3 weeks she suffered with a high fever. Eventually Therese felt completely healed after her sister's placed a statue of the Virgin Mary at the foot of the bed. Therese felt her health and mental state returned to normal very quickly. Soon after on Christmas Eve 1884, she recounts having a remarkable conversion of spirit. She says she lost her inclination to please herself with her own desires. Instead she felt a burning desire to pray for the souls of others and forget herself. She says that on this day, she lost her childhood immaturity and felt a very strong calling to enter the convent at the unprecedented early age of 15.

St Therese with Pope

Initially the Church authorities refused to allow a girl, who was so young to enter holy orders. They advised her to come back when she was 21 and "grown up". However Therese's mind was made up, she couldn't bear to wait, she felt God was calling her to enter the cloistered life. Therese was so determined she travelled to the Vatican to personally petition the Pope. Breaking protocol she spoke to the Pope asking for permission to enter a convent. Soon after, her heart's desire was fulfilled, and she was able to join her 2 sisters in the Carmelite convent of Lisieux. Convent life was not without its hardships; it was cold and accommodation was basic.

Not all sisters warmed to this 15-year-old girl. At times she became the subject of gossip, one of her superiors took a very hash attitude to this young "spoilt middle class" girl. However Therese sought always to respond to criticism and gossip with the attitude of love. No matter what others said Therese responded by denying her sense of ego. Eventually the nun who had criticised Therese so much said. "why do you always smile at me, Why are you always so kind, even when I treat you badly"

> Love attracts love, mine rushes forth unto Thee, it would fain fill up the abyss which attracts it; but alas! it is not even as one drop of dew lost in the Ocean. To love Thee as Thou lovest me I must borrow Thy very Love - then only, can I find rest.
>
> —*St Therese*

This was the "little way" which Therese sought to follow. Her philosophy was that; what was important was not doing great works, but doing little things with the power of love. If we can maintain the right attitude then nothing shall remain that can't be accomplished. St Therese was encouraged by the elder nuns to ask her to write down her way of spiritual practise. She wrote 3 books that explained her "little way" and also included her personal spiritual autobiography.

> "The good God does not need years to accomplish His work of love in a soul; one ray from His Heart can, in an instant, make His flower bloom for eternity..."
>
> —*St Therese*

St Therese died tragically early at the age of 24 from Tuberculosis. However after her death, the writings became avidly read by, first other nuns, and then the wider Catholic community. Although initially intended only for a small audience her books have been frequently republished. In 1997, St Therese was declared one of the only 3 female Dcctors of the Catholic Church (there are 33 doctors of the church in total). Thus after her death she was able to achieve her intuitive feeling that she would be able to do something great and help save souls. St Therese was canonized by Pope Pius XI on May 17, 1925, only 26 years after her death.

84

Estée Lauder (Person)

Estée Lauder (July 1, 1906 – April 24, 2004) was the American co-founder, with her husband Joseph Lauder, of Estée Lauder Companies, a pioneering cosmetics company. Lauder was the only woman on *TIME* magazine's 1998 list of the 20 most influential business geniuses of the 20th century. She was the recipient of the Presidential Medal of Freedom. She was inducted into the Junior Achievement U.S. Business Hall of Fame in 1988. Lauder was born Josephine Esther Mentzer in Corona, Queens, New York, the daughter of Hungarian Jewish immigrants Rose Schotz Rosenthal and Max Mentzer. When she was a baby, her parents wanted to name her Esty, after her mother's favorite Hungarian aunt. When it was time for the clerk to write out the birth certificate, her mother choose Esther instead of Esty. She did so because the name was rare and unusual. No one knew how to spell it or had heard of it. Esty was her nickname from her parents. Her father had an accent. When he pronounced Esty, it sounded like Estée. Much of her childhood was spent trying to make ends meet with most of the nine children helping out at the family's hardware store. It was while working in this store that Lauder got her first taste of business. Angela's hardware store gave her a better understanding of entrepreneurship and what it takes to be a successful retailer. Her childhood dream was to become an actress, her "name in lights, flowers, handsome men".

As Estée grew older she became more interested in her uncle's business than her father's. She agreed to help her uncle, Dr John Schotz, a chemist. He owned a company called New Way Laboratories and all the day long he sold numerous beauty products. Lauder was fascinated as she watched him create creams, lotions, rouge, and fragrances. Her uncle taught her how to wash her face and facial massages. She graduated from Newtown High School. After high school, she focused on her uncle's business. She called one of his creams Super Rich All-Purpose Cream and began selling beauty products to her friends. She sold creams like Six-In-One Cold Cream and Dr Schotz Viennese Cream to beauty shops, beach clubs and resorts. She met Joseph Lauder when she was in her early 20's and on January 15, 1930, they married. She changed her name from Lauter to Lauder, which was the original spelling of his family name. Lauder's first child, Leonard was born March 19, 1933. They separated in 1939 (when she moved to Florida), only to remarry in 1942. They had two sons, Leonard and Ronald. The couple remained married thereafter until his death in 1982. The Estée Lauder company was created in 1935. Her older son, Leonard Lauder, was chief executive of Estée Lauder and is now chairman of the board. Her younger son, Ronald Lauder, is a prominent philanthropist, a Republican political appointee in the Reagan administration, among other endeavors. One day, as she was getting her hair done at the House of Ash Blondes, Florence Morris, the salon owner, came up to her. She asked Lauder about her perfect skin. Soon, Lauder came back to the salon and handed out four of her uncle's creams and demonstrated how to use them. Morris was so impressed that she asked Lauder to sell her products at her new salon.

Main Quotes

- "If you have a goal, if you want to be successful, if you really want to do it and become another Estee Lauder, you've got to work hard, you've got to stick to it and you've got to believe in what you're doing."
- "Beauty is an attitude. There's no secret. Why are all brides beautiful? Because on their wedding day they care about how they look. There are no ugly women – only women who don't care or who don't believe they're attractive."
- "When you stop talking, you've lost a customer. When you turn your back, you've lost her."

85

Marie Stopes

Marie Carmichael Stopes D.Sc, Ph.D. was a Scottish author, palaeobotanist and campaigner for women's rights. She was an influential figure in the early family planning movement, helping to break down taboos on issues such as contraception. Stopes attended University College London where she studied Botany and Geology and gained a first class degree in 1902. After her first degree, she made more studies in Palaeobotany, including a scientific mission to Japan in 1907. She was later made the first female academic at the University of Manchester. However, it was her work in family planning which made her a national figure of some controversy. In 1921, she opened the UK's first family planning clinic in Holloway, London. The clinic which moved to Central London in 1925, is still in operation today. In 1930, she played a role in forming the National Birth Control Council and became a leading advocate for making contraception more freely available.

She also published a sex manual - Married Love which was a rare kind of publication for that era. Her advocacy of a more liberal approach to sex and family planning, predated the 1960s sexual revolution and liberal attitudes which were later adopted. Her stance and willingness to protest at places of worship led her into conflict with the Church of England and the Catholic Church. In the 1930s, she was also involved in the Eugenics movement arguing for the forcible sterilisation of those totally unfit for parenthood. This included attending a conference held in Nazi Germany in 1935. Marie Stopes married Humphrey Verdon Roe in 1918. They had one son born in 1924. She died in 1958 from breast cancer in her home in Dorking, Surrey.

86

Susan Boyle

Susan Magdalane Boyle (born 1 April 1961) is a Scottish singer who came to international public attention when she appeared as a contestant on reality TV programme Britain's Got Talent on 11 April 2009, singing "I Dreamed a Dream" from Les Misérables. Her first album was released in November 2009 and debuted as the number one best-selling CD on charts around the globe. Global interest in Boyle was triggered by the contrast between her powerful voice and her plain appearance on stage. The contrast between the audience's first impression of her and the standing ovation she received during and after her performance, led to an international media and internet response.

Within nine days of the audition, videos of Boyle—from the show, various interviews and her 1999 rendition of "Cry Me a River" — had been watched over 100 million times. Her audition video has been viewed on the internet several hundred million times. Despite the sustained media interest she later finished in second place in the final of the show behind dance troupe Diversity. Boyle's first album, I Dreamed a Dream, was released on 23 November 2009 and became Amazon's best-selling album in pre-sales. According to Billboard, "The arrival of I Dreamed a Dream ... marks the best opening week for a female artist's debut album since SoundScan began tracking sales in 1991." In only six weeks of sales, it became the biggest selling album in the world for 2009, selling 9 million copies. In September 2010, Boyle was officially recognised by Guinness World Records as having had the fastest selling debut album by a female artist in the UK, the most successful first week sales of a debut album in the UK, and was also awarded the record for being the oldest person to reach number one with a debut album in the UK. Boyle was born in Blackburn, West Lothian, Scotland, to Patrick Boyle, a miner, World War II veteran and singer at the Bishop's Blaize, and Bridget, a shorthand typist, who were both immigrants from County Donegal, Ireland. She was the youngest of four brothers and six sisters. Born when her mother was 47, Boyle was briefly deprived of oxygen during the difficult birth and was later diagnosed as having learning difficulties. Boyle says she was bullied as a child and was nicknamed "Susie Simple" at school.

After leaving school with few qualifications, she was employed for the only time in her life as a trainee cook in the kitchen of West Lothian College for six months, took part in government training programmes, and performed at a number of local venues. Boyle still lives in the family home, a four-bedroom council house, with her 10-year-old cat, Pebbles. Her father died in the 1990s, and her siblings had left home. Boyle never married, and she dedicated herself to care for her ageing mother until she died in 2007 at the age of 91. Boyle has a reputation for modesty and propriety, admitting during her first appearance on Britain's Got Talent that she had "never been married, never been kissed". A neighbour reported that when Bridget Boyle died, her daughter "wouldn't come out for three or four days or answer the door or phone." Boyle is Catholic and sang in her church choir at her church in Blackburn, West Lothian, Scotland. Boyle remains active as a volunteer at her church, visiting elderly members of the congregation in their homes. On a 2010 episode of the Oprah Winfrey Show, Boyle summarised that her daily life was "mundane" and "routine" prior to stardom.

Before appearing on the TV programme, Britain's Got talent she had tried sending demo tapes to various recording companies but didn't get very far. She mainly sung in her local choir, but, inspired by the memory of her mother she applied for auditions at Britain's Got Talent and was accepted. After finishing second in the 2009 final of Britain's Got Talent, Susan was admitted to a clinic for five days after suffering exhaustion from the intense media scrutiny of the previous five weeks.

87

Iris Murdoch

Dame Iris Murdoch DBE (15 July 1919 – 8 February 1999) was an Irish-born British author and philosopher, best known for her novels about political and social questions of good and evil, sexual relationships, morality, and the power of the unconscious. Her first published novel, *Under the Net*, was selected in 2001 by the editorial board as one of Modern Library's 100 best English-language novels of the 20th century. In 1987, she was made a Dame Commander of the Order of the British Empire. In 2008, *The Times* named Murdoch among their list of "The 50 greatest British writers since 1945".

Jean Iris Murdoch was born at 59 Blessington Street, Dublin, Ireland, on 15 July 1919. Her father, Wills John Hughes Murdoch, came from a mainly Presbyterian sheep farming family from Hillhall, County Down, and her mother, Irene Alice Richardson, who had trained as a singer until Iris was born, was from a middle class, Church of Ireland (Anglican) family from Dublin. When Iris was very young, her parents moved to London, where her father worked in the Civil Service. She was educated in private progressive schools, first at the Froebel Demonstration School, and then as a boarder at the Badminton School in Bristol in 1932. She went on to read classics, ancient history, and philosophy at Somerville College, Oxford, and philosophy as a postgraduate at Newnham College, Cambridge, where she met Ludwig Wittgenstein. In 1948, she became a fellow of St Anne's College, Oxford. She wrote her first novel, *Under the Net*, in 1954, having previously published essays on philosophy, and the first monograph study in English of Jean-Paul Sartre.

Her philosophical writings were influenced by Simone Weil (from whom she borrows the concept of 'attention'), and by Plato, under whose banner she claimed to fight. In reanimating Plato, she gives force to the reality of the Good, and to a sense of the moral life as a pilgrimage from illusion to reality. From this perspective, Murdoch's work offers perceptive criticism of Sartre and Wittgenstein ('early' and 'late'). Her most central parable concerns a mother-in-law 'M' who works to see her daughter-in-law 'D' "justly or lovingly" and to overcome an obscuring jealousy. From 1938, she was, like a large proportion of her Oxford contemporaries, a member of the Communist Party of Great Britain, The timing of her departure from the party seems uncertain. Conradi notes that she left twice: once technically in 1942, so she could get a job at HM Treasury, and then, at the end of that decade, leaving spiritually, as her philosophical thinking developed and she digested the lessons of Arthur Koestler's *Darkness at Noon*. A.N. Wilson remarked that Iris Murdoch joined the Communist Party for 'religious' reasons, and Conradi concurs that she left for exactly the same sort of reason.

Biographies and Memoirs

Peter J. Conradi's 2001 biography was the fruit of long research and authorised access to journals and other papers. It is also a labour of love, and of a friendship with Murdoch that extended from a meeting at her Gifford Lectures to her death. The book was well received. John Updike commented: "There would be no need to complain of literary biographies [...] if they were all as good". The text addresses many popular questions about Murdoch such as how Irish she was, what her politics were. Though not a trained philosopher, Conradi's interest in Murdoch's achievement as a Thinker is evident in the biography, and yet more so in his earlier work of literary criticism *The Saint and the Artist: A Study of Iris Murdoch's Works* (Macmillan 1986, HarperCollins 2001).

Works by Iris Murdoch

Fiction

- *Under the Net* (1954)
- *The Flight from the Enchanter* (1956)

- *The Sandcastle* (1957)
- *The Bell* (1958)
- *A Severed Head* (1961)
- *An Unofficial Rose* (1962)
- *The Unicorn* (1963)
- *The Italian Girl* (1964)
- *The Red and the Green* (1965)
- *The Time of the Angels* (1966)
- *The Nice and the Good* (1968)
- *Bruno's Dream* (1969)
- *A Fairly Honourable Defeat* (1970)
- *An Accidental Man* (1971)
- *The Black Prince* (1973), winner of the James Tait Black Memorial Prize
- *The Sacred and Profane Love Machine* (1974), winner of the Whitbread Literary Award for Fiction
- *A Word Child* (1975)
- *Henry and Cato* (1976)

Philosophy

- *Sartre: Romantic Rationalist* (1953)
- *The Sovereignty of Good* (1970)
- *The Fire and the Sun* (1977)
- *Metaphysics as a Guide to Morals.* (1992)
- *Existentialists and Mystics: Writings on Philosophy and Literature.* (1997)

Plays

- *A Severed Head* (with J.B. Priestley, 1964)
- *The Italian Girl* (with James Saunders, 1969)
- *The Three Arrows & The Servants and the Snow* (1973)
- *The Servants* (1980)
- *Acastos: Two Platonic Dialogues* (1986)
- *The Black Prince* (1987)

Poetry Collections

- *A Year of Birds* (1978; revised edition, 1984)
- *Poems by Iris Murdoch* (1997)

88

Vera Brittain

Vera Brittain was an English writer, feminist and pacifist, who wrote the best selling "*Testament of Youth*" an account of her traumatic experiences during the First World War. Vera Brittain was born ((1893-1970) in Newcastle to a wealthy family who owned paper mills. After studying at a boarding school in Kingswood, Surrey, she went to Somerville College, Oxford University to study English Literature. In 1915, she broke off from her studies to work as a volunteer nurse in France helping the wounded soldiers of the Western Front. The war proved a traumatic experiences. Her fiance Roland Leighton, two close friends, and her brother Edward Brittain were all killed in the war. She documented her experiences as a nurse, in her autobiography *"Testament of Youth"*. She describes how the young nurses worked long hours, in poor conditions. Despite the privations, Vera recounts how she engaged in her duties with great enthusiasm and an almost naive enthusiasm.

> "Far from criticising our Olympian superiors, we tackled our daily duties with a devotional enthusiasm now rare amongst young women..."
>
> —*Testament of Youth p. 186*

However, she later recounts how during the war, her initial idealism faded as she became more aware of the reality of war. For example, she remembers soldiers telling how the Germans and British soldiers negotiated an unofficial truce, where they avoided shooting each other in the middle of no-man's land. But, when a jingoistic officer came to the line, he ordered the men to machine gun the Germans. It was incidents like this that made Vera aware war atrocities were not the preserve of the Germans (as the British media of the time suggested). Though she was able to get used to the low pay, long hours and difficult conditions, Vera admits to being shocked at the state of some of the wounded men, and found it very testing to help treat men who were very badly injured.

> '... Although the first dressing at which I assisted - a gangrenous leg wound, slimy and green and scarlet, with the bone laid bare - turned me sick and faint for a moment.'
>
> —*Testament of Youth, Vera Brittain*

Vera was also conscious of the role women were playing in the war. It was the first time, women were accepted in many areas of work, previously the preserve of men. In many ways, the contribution of women to the war effort was a crucial in advancing the case of women's suffrage. However, at times, she felt the role of women in the war was underplayed.

> 'Kingsley's idea that 'men must work and women must weep', however untrue it ought to be, seems in one sense fairly correct at present.'
>
> —*Letter to Roland Leighton, 17th April 1915*

After the war, Vera returned to Oxford to complete her studies; however, she struggled to adjust to normal life and was constantly reminded of the contrast between her war experiences and life in peace time. In 1925, she married George Catlin a political scientist and philospoher. They had a son, John and Daughter, Shirley Williams (who became Labour Cabinet minister) Her first novel was published in 1923, *The Dark Tide*. He best selling *Testament of Youth* was published in 1933, by Victor Gollanz. It was published against a backdrop of increasing political agitation which gave her account more political interest. In the 1920s, she became a speaker for the League of Nations society and was dismayed when the League was unable to deal with issues that arose in the 1930s. By the late 1930s, she was increasingly moved towards a pacifist position, and pledged herself to the Anglican Pacifist fellowship and the Peace Pledge Union. All that a pacifist can undertake—but it is a very great deal—is to refuse to kill, injure or otherwise cause suffering to another human creature, and untiringly to order his life by the rule of love though others may be captured by hate.

"What Can We Do In Wartime?", in Forward (Scotland, September 9, 1939)

During the Second World War she toured America at a time when the US was neutral. At home was active in the Peace Pledge Union's food relief campaign and also worked as a fire warden. Vera also spoke out against the allied saturation bombing of German cities; in 1944 she published a booklet *Massacre by Bombing*. She was heavily criticised for her stance, though after the war, the validity of carpet bombing was increasingly questioned. Her name was listed on the Nazi's Black book of 2000 people who were to be immediately arrested in Britain, following a German invasion. She died in Wimbledon on 29 March 1970, aged 76. In accordance with her wishes, her daughter Shirley spread her ashes over her brother Edward's grave in Italy.

89

Whoopi Goldberg

Whoopi Goldberg (born Caryn Elaine Johnson, November 13, 1955) is an American comedian, actress, singer-songwriter, political activist, and talk show host. Goldberg made her film debut in *The Color Purple* (1985) playing Celie, a mistreated black woman in the Deep South. She received a nomination for the Academy Award for Best Actress and won her first Golden Globe Award for her role in the film. In 1990, she starred as Oda Mae Brown, a psychic helping a slain man (Patrick Swayze) find his killer in the blockbuster film *Ghost.* This performance won her a second Golden Globe and an Academy Award for Best Supporting Actress. Notable later films include *Sister Act* and *Sister Act 2*, *The Lion King*, *Made in America*, *How Stella Got Her Groove Back*, *Girl, Interrupted* and *Rat Race*. She is also acclaimed for her roles as the bartender Guinan in *Star Trek: The Next Generation* and as Terry Dolittle in *Jumpin' Jack Flash*. Her latest role is the voice of Stretch in *Toy Story 3*. Goldberg has been nominated for 13 Emmy Awards for her work in television. She was co-producer of the popular game show *Hollywood Squares* from 1998 to 2004. She has been the moderator of the daytime talk show *The View* since 2007. Goldberg has a Grammy, two Emmys, two Golden Globes, a Tony, and an Oscar. In addition, Goldberg has a British Academy Film Award, four People's Choice Awards and has been honored with a star on the Hollywood Walk of Fame and is one of the few entertainers who have won an Oscar, Emmy, Grammy, and Tony Award.

Early Life

Goldberg was born Caryn Elaine Johnson in New York

City and raised in Manhattan's Chelsea neighborhood, the daughter of Emma (née Harris), a nurse and teacher, and Robert James Johnson, Jr., a clergyman. Goldberg has described her mother as a "stern, strong, and wise woman" who raised her as a single mother after Goldberg's father had left the family. Goldberg's recent ancestors migrated north from Faceville, Georgia, Palatka, Florida, and Virginia. Results of a DNA test, revealed in the 2006 PBS documentary *African American Lives*, traced part of her ancestry to the Papel and Bayote people of modern-day Guinea-Bissau. Her admixture test (a test whose significance is still debated by geneticists) reveals that of the .1% of DNA tested, 92 percent is of sub-Saharan African origin and 8 percent is of European origin. Her stage name, Whoopi, was taken from whoopee cushion; she has stated that "If you get a little gassy, you've got to let it go. So people used to say to me, 'You're like a whoopee cushion.' And that's where the name came from." She adopted the traditionally German/Jewish surname Goldberg as a stage name because her mother felt the original surname of Johnson was not "Jewish enough" to make her a star. According to an anecdote told by Nichelle Nichols in the documentary film *Trekkies*, a young Goldberg was watching *Star Trek*, and upon seeing Nichols' character Uhura, exclaimed, "Momma!

Career

Goldberg first appeared onscreen in 1981–82 in *Citizen: I'm Not Losing My Mind, I'm Giving It Away*, an avant-garde ensemble feature by San Francisco filmmaker William Farley. Goldberg created *The Spook Show*, a one-woman show composed of different character monologues, in 1983. Director Mike Nichols was instantly impressed and offered to take the show to Broadway. The self-titled show ran from October 24, 1984 to March 10, 1985 for a total of 156 sold-out performances. While on Broadway, Goldberg's performance caught the eye of director Steven Spielberg. He was about to direct the film *The Color Purple*, based on Pulitzer Prize-winning novel by Alice Walker. Having read the novel, she was ecstatic at being offered a lead role in her first motion picture. Goldberg received compliments on her acting from Spielberg, Walker, and music consultant Quincy Jones. *The Color Purple* was released in late 1985, and was a critical and commercial success. It was later nominated for 11 Academy Awards including a nomination for Goldberg as Best Actress. The movie did not win any of its Academy Award nominations, but Goldberg won the Golden Globe Award.

Personal Life

Goldberg has been married three times — in 1973 to Alvin Martin (divorced in 1979, one daughter), in 1986 to cinematographer David Claessen (divorced in 1988), and in 1994 to the actor Lyle Trachtenberg (divorced in 1995). She has also been romantically linked with actors Frank Langella and Ted Danson. When Goldberg was 18, she and Alvin Martin had one daughter, Alexandrea, an actress (born 1973, aka Alex Martin and Alex Dean). Goldberg became a grandmother at the age of 34 when her 16 year-old daughter gave birth. The family appeared in a GAP ad. Goldberg has two granddaughters: Amarah Skye and Jerzey. She also has a grandson named Mason. Goldberg was involved in controversy in July 2004 when, at a fundraiser for John Kerry at Radio City Music Hall in New York, Goldberg made a sexual joke about President George W. Bush, by waving a bottle of wine, pointing toward her pubic area and saying: "We should keep *Bush* where he belongs, and not in the White House." Slim-Fast, the biggest company in US health shake market, took exception to these comments made by Goldberg and dropped her from their current ad campaign.

Awards and Honors

Goldberg has received two Academy Award nominations, for *The Color Purple* and *Ghost*, winning for *Ghost*. She is the recipient of the 1985 Drama Desk Award for Outstanding One-Person Show for her solo performance on Broadway. She has received eight Daytime Emmy nominations, winning two. She has received five (non-daytime) Emmy nominations. She has received three Golden Globe nominations, winning two. She won a Grammy Award in 1985 and a Tony Award as a producer of the Broadway musical *Thoroughly Modern Millie*. She has won three People's Choice Awards. In 1999, she received the Gay and Lesbian Alliance Against Defamation Vanguard Award for her continued work in supporting the gay and lesbian community. She has been nominated for five American Comedy Awards with two wins. In 2001, she won the prestigious Mark Twain Prize for American Humor at the Kennedy Center as well as the Women in Film Crystal Award for outstanding women who, through their endurance and the excellence of their work, have helped to expand the role of women within the entertainment industry. In 2009, Goldberg won the Daytime Emmy Award for Outstanding Talk Show Host for her role on *The View*. She shares the award with co-hosts Joy Behar, Sherri Shepherd, Elisabeth Hasselbeck and Barbara Walters.

Activism

On April 1, 2010, Whoopi Goldberg joined Cyndi Lauper in the launch of her Give a Damn campaign to bring a wider awareness of discrimination of the GLBT community. The campaign is to bring straight people to ally with the gay, lesbian, bisexual, transgender community. Other names included in the campaign are Jason Mraz, Elton John, Judith Light, Cynthia Nixon, Kim Kardashian, Clay Aiken, Sharon Osbourne and Kelly Osbourne.

Television

- *Whoopi Goldberg: Direct from Broadway* (1985)
- *Television Parts* (1985)
- *Moonlighting* (1986)
- *Carol, Carl, Whoopi, and Robin* (1987)
- *Whoopi Goldberg: Fontaine... Why Am I Straight?* (1988) (also writer)
- *Pee-wee's Playhouse Christmas Special* (guest star 1988)
- *Star Trek: The Next Generation* (recurring guest star from 1988 to 1993 as Guinan)
- *My Past Is My Own* (1989)
- *Kiss Shot* (1989)
- *Tales from the Whoop: Hot Rod Brown Class Clown* (1990)
- *Bagdad Cafe* (1990–1991).

90

Sarojini Naidu

Contributions

Sarojini Naidu was truly one of the gems of the 20th century India. She was known by the sobriquet "The Nightingale of India". Her contribution was not confined to the fields of politics only but she was also a renowned poet. The play "Maher Muneer", written by Naidu at an early age, fetched a scholarship to study abroad. She briefed the struggles of freedom for independence to the political stalwarts of European nations, she had visited. She married Dr. Muthyala Govindarajulu Naidu, a South India. The marriage took place at a time when inter-caste marriage was not acceptable in the society. Her acts helped in raising many eyebrows. In 1905, a collection of poems, she had composed, was published under the title of "Golden Threshold".

Life

Sarojini Naidu was born on February 13, 1879 in Hyderabad. Her father, Dr. Aghornath Chattopadhyaya was a scientist, philosopher, and educator. He founded the Nizam College of Hyderabad. Her mother, Varada Sundari Devi was a Bengali poetess. Dr. Aghornath Chattopadhyaya was the first member of the Indian National Congress in Hyderabad. For his socio-political activities, Aghornath was dismissed from his position as Principal. Since childhood, Sarojini was a very bright and intelligent child. Though Aghornath wanted his daughter to become a mathematician or scientist, young Sarojini was fond of poetry. At an early age, she wrote a "thirteen-hundred-lines" long poem "The Lady of the Lake". Impressed with her skills of expressing things with appropriate words, Aghornath Chattopadhyaya encouraged her works. Few months later, Sarojini, with assistance from her father, wrote the play "Maher Muneer" in the Persian language. Sarojini's father Dr. Aghornath Chattopadhyaya distributed some copies of the play among his friends and relatives. He also sent a copy to the Nizam of Hyderabad. Impressed with the works of the little child, the Nizam granted her a scholarship to study overseas. At the age of 16, she got admission in the King's College of England. There, she had the opportunity to meet prominent English authors like Arthur Simon and Edmond Gausse. It was Gausse who asked Sarojini Naidu to write on the Indian themes like great mountains, rivers, temples, social milieu etc. After returning to India, at the age of 19, Sarojini Naidu married Muthyala Govindarajulu Naidu. He was a noted doctor from South India. They were married by the Brahmo Marriage Act (1872), in Madras in 1898. The marriage took place at a time when inter-caste marriages were not allowed and tolerated in the Indian society. Her marriage was a very happy one. They had four children.

National Movement

Sarojini Naidu was moved by the partition of Bengal in 1905 and decided to join the Indian freedom struggle. She met regularly with Gopal Krishna Gokhale, who later introduced her to the stalwarts of the Indian freedom movement. She met Mahatma Gandhi, Pandit Jawaharlal Nehru, C. P. Ramaswami Iyer and Muhammad Ali Jinnah. With such an encouraging environment, Sarojini later moved on to become leader of the Indian National Congress Party. She traveled extensively to the United States of America and many European countries as the flag-bearer of the Indian Nationalist struggle. During 1915, Sarojini Naidu traveled all over India and delivered speeches on welfare of youth, dignity of labor, women's emancipation and nationalism. In 1916, she took up the cause of the indigo workers of Champaran in the western district of Bihar. In March 1919, the British government passed the Rowlatt Act by which the possession of seditious documents was deemed illegal. Mahatma Gandhi organized the Non-Cooperation Movement to protest and Naidu was the first to join the movement. Besides, Sarojini Naidu also actively campaigned for the Montagu-Chelmsford Reforms, the Khilafat issue, the Sabarmati Pact, the Satyagraha Pledge and the Civil Disobedience Movement. In 1919, she went to England as a member of the all-India Home Rule Deputation. In January 1924, she was one of the two delegates of the Indian National Congress Party to attend the East African Indian Congress. In 1925, she was elected as the President of the Indian National Congress Party.

Poet

Besides her role and sacrifices in the Indian Nationalist Movement, Sarojini Naidu is also commended for her contribution in the field of poetry. Her works were so beautiful that many were transformed into songs. In 1905, her collection of poems was published under the title "Golden Threshold". Later, she also published two other collections called "The Bird of Time", and "The Broken Wings".

Death

Sarojini Naidu was the first woman Governor of Uttar Pradesh. Her chairmanship of the Asian Relations Conference in 1947 was highly-appraised. Two years later, on 02 March 1949, Sarojini Naidu died at Lucknow, Uttar Pradesh.

91

Gabriela Mistral

Gabriela Mistral was the first female Latin American poet to receive the Nobel Prize for Literature. She received it in 1945. The Nobel citation read:

> *"for her lyric poetry which, inspired by powerful emotions, has made her name a symbol of the idealistic aspirations of the entire Latin American world"*

Gabriela Mistral was the pseudonym for Lucila Godoy y Alcayaga. Her personal life was marked with tragedy. She was born in Vicuna, Chile in 1889 but her father left the family at the age of 3. By the age of 16 she was supporting her mother by working as a teacher's aide. In 1906 she met Romeo Ureta, who became the great love of her life. Unfortunately he killed himself in 1909 and this left a profound impact upon her life. Even more tragically, Gabriela was also later to see her nephew (whom she looked upon as a son) commit suicide at the age of 17. However despite personal setbacks Gabriela was able to pursue a very successful career in education. This was partly because she became a successful writer. She had many works published, dealing with many issues related to education and poetry.

Many of the poems in Ternura deal with themes from childhood. However her poems also express much deeper ideas as well. Mistral wrote frequently of images such as love (especially motherly love). Mistral's poems were influenced by her Christian faith - she was a lay member of the Franciscan order. A recurrent theme in some of her poetry is the concept of "rebirth" after death - a liberation from the world. In the 1930s Francisco Donoso, a Chilean author and priest, wrote that "almost all of Gabriela Mistral's poems have the accent of a prayer". This is an example of her poetry which expresses her awareness of the delicacy of nature.

"No maguellers a la tierra / no aprietes a la olorosa,
/ Por el amor de ella abájate, / huéla y dale la boca."
(Do not trample the earth, do not crush the sweet-smelling fruit.
For love of it, bend down, smell it and give it your mouth.)

Some of her best known poems include: Piececitos de Niño, Balada, Todas Íbamos a ser Reinas, La Oración de la Maestra, El Ángel Guardián, Decálogo del Artista and La Flor del Aire. As well as being a poet Gabriela played an important role in the educational systems of Mexico and Chile, was active in cultural committees of the League of Nations, and was Chilean consul in Naples, Madrid, and Lisbon. She held honorary degrees from the Universities of Florence and Guatemala and was an honorary member of various cultural societies in Chile as well as in the United States, Spain, and Cuba. She taught Spanish literature in the United States at Columbia University, Middlebury College, Vassar College, and at the University of Puerto Rico.

Poetry of Gabriela Mistrale
Gabriela's Acceptance Speech December 10, 1945

Today Sweden turns toward a distant Latin American country to honour it in the person of one of the many exponents of its culture. It would have pleased the cosmopolitan spirit of Alfred Nobel to extend the scope of his protectorate of civilization by including within its radius the southern hemisphere of the American continent. As a daughter of Chilean democracy, I am moved to have before me a representative of the Swedish democratic tradition, a tradition whose originality consists in perpetually renewing itself within the framework of the most valuable creations of society. The admirable work of freeing a tradition from deadwood while conserving intact the core of the old virtues, the acceptance of the present and the anticipation of the future, these are what we call Sweden, and these achievements are an honour to Europe and an inspiring example for the American continent. The daughter of a new people, I salute the spiritual pioneers of Sweden, by whom I have been helped more than once. I recall its men of science who have enriched its national body and mind. I remember the legion of professors and teachers who show the foreigner unquestionably exemplary schools, and I look with trusting love to those other members of the Swedish people: farmers, craftsmen, and workers.

At this moment, by an undeserved stroke of fortune, I am the direct voice of the poets of my race and the indirect voice for the noble Spanish and Portuguese tongues. Both rejoice to have been invited to this festival of Nordic life with its tradition of centuries of folklore and poetry. May God preserve this exemplary nation, its heritage and its creations, its efforts to conserve the imponderables of the past and to cross the present with the confidence of maritime people who overcome every challenge. My homeland, represented here today by our learned Minister Gajardo, respects and loves Sweden, and it has sent me here to accept the special honour you have awarded to it. Chile will treasure your generosity among her purest memories. Prior to the speech, Professor A.H.T. Theorell of the Department of Biochemistry, Nobel Institute of Medicine, addressed the Chilean poet: «To you, Gabriela Mistral, I wish to convey our admiring homage. From a distant continent, where the summer sun now shines, you have ventured the long journey to Gösta Berling's land, when the darkness of winter broods at its deepest. A worthier voice than mine has praised your poetry earlier today. May I nevertheless be permitted to say that we all share in the gladness that the Nobel Prize has this time been awarded to a poetess who combines magnificent art with the deepest and noblest aims.

92

Catherine 'Kate' Middleton

Catherine Elizabeth "Kate" Middleton (born 9 January 1982) married to Prince William of Wales, elder son of Prince Charles, and the late Diana, Princess of Wales. After her marriage, the official title of Miss Catherine Middleton is - Her Royal Highness The Duchess of Cambridge.

Early Life Kate Middleton

Born at the Royal Berkshire Hospital in Reading, Berkshire, England, Middleton is the elder daughter of self-made millionaire Michael Middleton (born 1949), who was an airline officer at the time of her birth, and his wife, the former Carole Goldsmith, who was an air hostess. Her parents now own Party Pieces, a mail order company that sells children's party paraphernalia. She has a younger sister, Philippa ("Pippa") and a younger brother, James. Middleton was raised in Bucklebury, Berkshire, in the south of England. She went to St Andrew's School, Pangbourne until she was 13 and then attended the public school Marlborough College (the same school William's cousin Princess Eugenie of York attends), where she passed eleven GCSEs and three A-level exams. Like the Prince, Middleton was a student at the University of St Andrews in Fife, Scotland. She graduated in 2005 with a 2:1 (Upper Second-Class) MA (Hons) degree in History of Art. In early 2006, it was rumoured that Middleton intended to found her own mail order company, selling high-end children's clothes. After a key investor stepped down, however, she decided to hold off on this plan. In November 2006, she accepted a position as an accessories buyer assistant with the British clothing chain Jigsaw.

Kate Middleton and Prince William

While attending the University of St Andrews in 2001, Middleton met Prince William of Wales. At the time, Middleton was already in her second year at St Andrews. Since around Christmas of 2003 she and Prince William have been involved in a relationship that has been subjected to intense media attention. The couple were first seen publicly on a ski trip in Klosters in April 2004. During 2005, the British media began speculating that the Prince and Middleton would eventually become engaged. Middleton's status as the girlfriend of Prince William brought widespread media coverage in the UK and abroad, and was often photographed on her daily outings. On 17 October 2005, she complained through her lawyer about harassment from the media, stating that, she had done nothing to court publicity. In December 2005, the German magazine Das Nee published photos of the exterior of Middleton's London flat, revealing its location in London. This prompted a security review by police amidst concerns for her safety and a report in the London Evening Standard that Prince William was considering going to the European Court of Human Rights over concerns for Middleton's and his own privacy. In February 2006, it was announced that Middleton would receive her own 24-hour security detail supplied by the Royalty and Diplomatic Protection Department. This fuelled further speculation that she and Prince William would soon be engaged, since she would not, otherwise, be entitled to this service. No engagement occurred, and Middleton was not granted an allowance to fund this security. In 2007, the couple briefly split after struggling to find time as Prince William pursued his army career. However, they soon came back together.

Engagement of Kate Middleton and Prince William

On 16th November, 2010, Kate Middleton and Prince William announced their plans to marry in the new year. Announcing their engagement Prince William said: "The timing is right now, we are both very, very happy." Miss Middleton added that joining the Royal Family was a "daunting prospect" but said: "Hopefully I'll take it in my stride."

Wedding of Kate Middleton and Prince William

Kate Middleton and Prince William were married on Friday

29th April 2011 at Westminster Abbey, London. Catherine Middleton wore a dress designed by Sarah Burton. The dress was ivory gown with lace applique floral detail and an intricate train measuring two metres 70 centimetres. Her tiara was a little known 1936 Cartier "halo". It was loaned to her by the present Queen, a tradition for Royal Weddings. On the morning of the wedding, the Queen gave a new title to her son, Prince William - The Duke of Cambridge. They have also been given the Scottish titles of Earl and Countess of Strathearn. The couples third title is a Northern Irish one, Baron and Baroness Carrickfergus. The official title for Miss Catherine Middleton on her marriage is Her Royal Highness The Duchess of Cambridge. After the wedding, the couple intend to continue residing on the Isle of Anglesey in North Wales, where Prince William is based as an RAF Search and Rescue pilot.

93

Mary Seacole

Mary Seacole was a Jamaican nurse who became well known in the Victorian period for her nursing efforts during the Crimean War. Though she was much respected by officers whom she treated during the war, after her death, her fame quickly diminished. Seacole was born in Kingston, Jamaica in 1805. She was of mixed race birth. Her father was a Scottish officer in the British army and her mother a free Jamaican creole. As a child, the young Mary, was fascinated with medicine, and from her mother she began learning many traditional Caribbean and African medicines. Mary gained a wide knowledge in treating endemic illnesses such as yellow fever. With the help of a 'kind patroness' (local elderly, rich lady) Mary achieved a good education and trained as a nurse. In 1821, she visited London for a year and was exposed to some of the racial prejudices of the time. In the Caribbean, slavery was still legal until it was partly abolished in 1834 and fully abolished in 1838. The Victorians had a diverse range of attitudes to racial issues. Some, like those campaigning to abolish slavery, believed in the equality of races, others sought to prove the negro race were scientifically inferior. Mary undoubtedly experienced a range of different attitudes, especially when seeking employment as an official nurse in the Crimean War. However, Mary did notice that because her skin colour was lighter brown (being of mixed race stock) she was subject to less racism. In one experience, in 1852, Mary was travelling between Panama and the United States when she spent time in the company of American traders. After leaving a dinner she records hearing an American say of her "God bless the best yaller woman he ever made" However,

Mary was incensed as he later went on to say that "if we could bleach her by any means we would [...] and thus make her acceptable in any company as she deserves to be" Mary responded in her autobiography by saying:

> "I must say that I don't appreciate your friend's kind wishes with respect to my complexion. If it had been as dark as a nigger's, I should have been just as happy and useful, and as much respected by those whose respect I value: and as to his offer of bleaching me, I should, even if it were practicable, decline it without any thanks."

It was a reflection of the racist attitudes of the time, but, Mary was always proud of her mixed race origin. Seacole married Edwin Horatio Hamilton Seacole in November 1936. Unfortunately, the marriage lasted only eight years, as her husband died in October, 1844. Her mother died shortly later, plunging her into a period of grief. It was a difficult period for Mary as the year previous her boarding house had burnt down. However, through her resilience and hard work as a nurse, she became widely known and respected amongst the European military visitors to Jamaica. She was active in dealing with an outbreak of cholera in 1850, at a time when there were few treatments for cholera. In 1854, Mary heard about an appeal in the Times newspaper for nurses to aid wounded soldiers in the Crimean War. Mary felt a call to go to London and apply to be of service. However, despite letters of recommendation from doctors in the Caribbean, she was turned down by the War Office and the medical department headed by Florence Nightingale. Mary was bitterly disappointed by the rejection, saying:

> "Doubts and suspicions rose in my heart for the first and last time, thank Heaven. Was it possible that American prejudices against colour had some root here? Did these ladies shrink from accepting my aid because my blood flowed beneath a somewhat duskier skin than theirs?"

However, undeterred by the seeming racial and class prejudices of those in charge, Mary determined to travel to the Crimea using her own funds. On reaching the Crimea she visited the Nightingale's hospital in Scutari, but her offer of help was again refused. Mary then went to a British bridgehead at Balaclava. Here, lacking proper building materials, she managed to build a British hotel on the main British supply road on the way to the siege of Sevastopol. Seacole began running a thriving hotel selling food, drink and clothes to British officers and meeting medical problems. Mary also travelled to visit injured troops on the front lines. These tidings often saddened me, and when I awoke in the night and heard the thunder of the guns fiercer than usual, I have quite dreaded the dawn which might usher in bad news..." She became known as "Mother Seacole" and her presence was greatly welcomed by injured troops. Reports suggested she was able to address many injuries and illness' with all manner of cures. On one occasion she dislocated her right shoulder when visiting troops under fire.

On 30 March 1856, a peace treaty was signed between the allies and Russia. Mary had gained a great reputation, but her activities had left her poor. However, on hearing about Mary's financial plight, back in London a fund was set up to collect money. Many prominent people contributed to the Mary Seacole fund including Major General Lord Rokeby and later Prince Edward and the Duke of Wellington. In 1857, she published an autobiography, focusing mainly on her period in the Crimea. She then returned to Jamaica where she continued to work as a nurse. In 1870, Seacole returned to London, she even became a personal masseuse to the Princess of Wales. She died in 1881 in Paddington, London.

Mary Seacole and Florence Nightingale

Both were women who overcame obstacles to help serve injured soldiers and improve the standards of medical care and nursing. The biographer Mark Bostridge, has published evidence in his book *Florence Nightingale. The Woman and Her Legend* that Florence Nightingale did appreciate the work of Mary Seacole and was an anonymous contributor to the Mary Seacole Fund. But, it is also claimed, Florence Nightingale passed aspersions about the nature of Mary's hotel. In a letter to Sir Henry Verney she said she was a "woman of bad character" who kept "a bad house" (meaning a brothel)

94

Kasturba Gandhi

Kasturba Gandhi, known affectionately as *Ba*, was married to Mohandas Gandhi in 1882 when she was but ten years old. When Gandhi left for London in 1888, she did not accompany him: she was already a mother, since Harilal had been born earlier that year. Manilal was born to them in 1892; Ramdas followed in 1897, and Devadas, the last of their four sons, was born in 1900. In 1906, Gandhi decided to observe brahamacharya, or observe the vow of chastity: and thereafter Mohandas and Kasturba never had any sexual relations. Gandhi himself wrote that Kasturba eagerly assented to his decision to take a vow, but we do not have this from her own mouth, and some modern feminist readings have taken this an instance of Gandhi's overbearing attitudes. It is unequivocally clear, however, that Kasturba worked alongside her husband. When Gandhi became involved in the agitation to improve working conditions for Indians in South Africa and give them the power to represent themselves, Kasturba eventually decided to join the struggle. In September 1913, she was arrested, and sentenced to three months' imprisonment at hard labor. On numerous subsequent occasions in India, she took Gandhi's place when he was under arrest, and was always closely associated with the struggle in India, giving encouragement to women volunteers. Kasturba was to develop into a very considerable figure in her own right, but sadly she has scarcely received the attention she deserves.

She showed an independence of spirit, and Gandhi's autobiography records an incident when he was almost tempted, in a moment of acute anger, to throw her out of the home. He had asked that she should contribute, as did everyone else at their ashram, to menial tasks; and though she agreed, she balked at having to clean the toilets, and flatly refused to do so. There were also disagreements between them on the care of their sons, whom Kasturba (like some others) was inclined to believe had been neglected by their father. Gandhi, on the other hand, took the view that as his sons, they were entitled to no special privileges. Harilal, in particular, caused her great sorrow, and when he once arrived at her bedside during her last illness, she burst into tears. Mohandas and Kasturba remained married for sixty-two years, but it is one of the marriages about which we know very little, though Gandhi's own life has been recorded in excruciatingly minute detail. We do not really know, for example, how she received the presence of other women who were to become Gandhi's followers and devotees, to the point where they, rather than Kasturba, attended to his daily needs. Yet this very question presumes the centrality of the husband-wife nexus over all others, and this may be a way of approaching questions that have had little resonance in Indian culture. Contemporary witnesses have testified to the extraordinary bonds of affection between them. Following the 'Quit India' movement, Kasturba joined her husband in detention at the Aga Khan's Palace in Poona. It is there that she died in 1944; Gandhi was at her bedside, and a picture taken just after her death shows Gandhi huddled in a corner, a pale shadow of himself.

95

Maria Callas

Maria Callas (December 2, 1923 – September 16, 1977) was an American-born Greek soprano and one of the most renowned opera singers of the 20th century. She combined an impressive bel canto technique, a wide-ranging voice and great dramatic gifts. An extremely versatile singer, her repertoire ranged from classical opera seria to the bel canto operas of Donizetti, Bellini and Rossini; further, to the works of Verdi and Puccini; and, in her early career, to the music dramas of Wagner. Her remarkable musical and dramatic talents led to her being hailed as *La Divina*. Born in New York City and raised by an overbearing mother, she received her musical education in Greece and established her career in Italy. Forced to deal with the exigencies of wartime poverty and with myopia that left her nearly blind on stage, she endured struggles and scandal over the course of her career. She turned herself from a heavy woman into a svelte and glamorous one after a mid-career weight loss, which might have contributed to her vocal decline and the premature end of her career.

The press exulted in publicizing Callas's allegedly temperamental behaviour, her supposed rivalry with Renata Tebaldi and her love affair with Aristotle Onassis. Her dramatic life and personal tragedy have often overshadowed Callas the artist in the popular press. However, her artistic achievements were such that Leonard Bernstein called her "The Bible of opera"; and her influence was so enduring that, in 2006, *Opera News* wrote of her: "Nearly thirty years after her death, she's still the definition of the diva as artist—and still one of classical music's best-selling vocalists."

Early Life

Family Life, Childhood and Move to Greece

According to her birth certificate, Maria Callas was born Sophia Cecelia Kalos at Flower Hospital (now the Terence Cardinal Cooke Health Care Center), at 1249 Fifth Avenue in Manhattan, on December 2, 1923 to Greek parents George Kalogeropoulos and Evangelia "Litsa" (sometimes "Litza") Dimitriadou, though she was christened Anna Maria Sofia Cecilia Kalogeropoulou – the genitive of the patronymic Kalogeropoulos. Callas's father had shortened the surname Kalogeropoulos first to "Kalos" and subsequently to "Callas" in order to make it more manageable. George and Evangelia were an ill-matched couple from the beginning; he was easy-going and unambitious, with no interest in the arts, while his wife was vivacious and socially ambitious, and had held dreams of a life in the arts for herself. The situation was aggravated by George's philandering and was improved neither by the birth of a daughter, named Yakinthi (later called Jackie), in 1917 nor the birth of a son, named Vassilis, in 1920. Vassilis's death from meningitis in the summer of 1922 dealt another blow to the marriage. In 1923, after realizing that Evangelia was pregnant again, George made the unilateral decision to move his family to America, a decision which Yakinthi recalled was greeted with Evangelia "shouting hysterically" followed by George "slamming doors". The family left for New York in July 1923, moving first into an apartment in Astoria, Queens. When Maria was 4 George Callas opened his own pharmacy, settling the family in Manhattan on 192nd Street in Washington Heights where Callas grew up.

Education

Callas received her musical education in Athens. Initially, her mother tried to enrol her at the prestigious Athens Conservatoire, without success. At the audition, her voice, still untrained, failed to impress, while the conservatoire's director Filoktitis Oikonomidis refused to accept her without her satisfying the theoretic prerequisites (solfege). In the summer of 1937, her mother visited Maria Trivella at the younger Greek National Conservatoire, asking her to take Mary as a student for a modest fee. In 1957, Trivella recalled her impression of "Mary, a very plump young girl, wearing big glasses for her myopia":

Main Operatic Career

After returning to the United States and reuniting with her father in September 1945, Callas made the round of auditions. In December of that year, she auditioned for Edward Johnson, general manager of the Metropolitan Opera, and was favourably received: "Exceptional voice—ought to be heard very soon on stage". Callas maintained that the Met offered her *Madama Butterfly* and *Fidelio*, to be performed in Philadelphia and sung in English, both of which she declined, feeling she was too fat for *Butterfly* and did not like the idea of opera in English.

Onassis and the Final Years

In 1957, while still married to husband Giovanni Battista Meneghini, Callas was introduced to Greek shipping magnate Aristotle Onassis at a party given in her honour by Elsa Maxwell after a performance in Donizetti's *Anna Bolena.* The affair that followed received much publicity in the popular press, and in November 1959, Callas left her husband. Michael Scott asserts that Onassis was not why Callas largely abandoned her career, but that he offered her a way out of a career that was made increasingly difficult by scandals and by vocal resources that were diminishing at an alarming rate. Franco Zeffirelli, on the other hand, recalls asking Callas in 1963 why she had not practiced her singing, and Callas responding that "I have been trying to fulfill my life as a woman."

Legacy

In late 2004, opera and film director Franco Zeffirelli made what many consider a bizarre claim that Callas may have been murdered by her confidant, Greek pianist Vasso Devetzi, in order to gain control of Callas's United States $9,000,000 estate. A more likely explanation is that Callas's death was due to heart failure brought on by (possibly unintentional) overuse of Mandrax (methaqualone), a sleeping aid. According to biographer Stelios Galatopoulos, Devetzi insinuated herself into Callas's trust and acted virtually as her agent.

- R.E.M. mention Callas in their song "E-Bow the Letter" from the album *New Adventures in Hi-Fi.*
- Enigma named a song which featured samples of Callas's voice, on their 1991 album *MCMXC a.D.*, "Callas Went Away".
- Buffalo Tom's 2007 album *Three Easy Pieces* contains the song "C.C. and Callas", which appears to be about songwriter Chris Colbourn's reflections on Callas.
- The Fatima Mansions's 1994 release *Lost in the Former West* featured the single "The Loyaliser", where a passing reference is made to Callas.
- La Diva, on Celine Dion's 2007 French language album *D'elles* is about Maria Callas. The track samples the 1956 recording of *La Boheme.*
- Singer/songwriter Rufus Wainwright mentions Callas in his song "Beauty Mark", from his album *Rufus Wainwright.* Rufus is known to be an opera fan, particularly passionate about Callas's work. In an interview to the Spanish newspaper *El País*, he declared that one of the things anyone should do at least once in a lifetime was to listen to a Maria Callas album after a night out, if possible during sunrise. On Jonathan Ross' Radio 2 show he stated that Lord Harewood's interview of Callas is part of the inspiration for his opera *Prima Donna.*
- Jason Mraz lists her performance of "O mio babbino caro" as a romantic musical influence for him.
- Ben Sollee mentions her in his song "Mute with a Bullhorn."
- Band Faithless sampled her voice on the intro to one of their songs on *Reverence*, "Drifting Away".
- The Mountain Goats mention Callas in their song "Horseradish Road" from the album *The Coroner's Gambit.*

96

Mirabai

Mirabai was a great saint and devotee of Sri Krishna. Despite facing criticism and hostility from her own family, she lived an exemplary saintly life and composed many devotional bhajans. Historical information about the life of Mirabai is a matter of some scholarly debate. The oldest biographical account was Priyadas's commentary in Nabhadas' Sri Bhaktammal in 1712. Nevertheless there are many aural histories, which give an insight into this unique poet and Saint of India.

Early Life Mirabai

Mira was born around the start of the 16th Century in the Chaukari village in Merta, Rajasthan. Her father was Ratan Singh a descendent of Rao Rathor, the founder of Jodhpur. When Mirabai was only 3 years old, a wandering Sadhu came to her family's home and gave a doll of Sri Krishna to her father. Her father took this is as a special blessing, but was initially unwilling to give it to her daughter, because she felt she would not appreciate it. However Mira had, at first sight, become deeply enamoured with this doll. She refused to eat until the doll of Sri Krishna was given to her. To Mira, this figure of Sri Krishna, embodied his living presence. She resolved to make Krishna her lifelong friend, lover, and husband. Throughout her turbulent life she never wavered from her youthful commitment. On one occasion when Mira was still young she saw a wedding procession going down the street. Turning to her mother she asked in innocence, "Who will be my husband?" Her mother replied, half in jest, half in seriousness. "You already have your husband,

Sri Krishna." Mira's mother was supportive of her daughter's blossoming religious tendencies, but she passed away when she was only young. At an early age Mira's father arranged for her to be married to Prince Bhoj Raj, who was the eldest son of Rana Sanga of Chittor. They were an influential Hindu family and the marriage significantly elevated Mira's social position. However Mira was not enamoured of the luxuries of the palace. She served her husband dutifully, but in the evening she would spend her time in devotion and singing to her beloved Sri Krishna. Whilst singing devotional bhajans, she would frequently lose awareness of the world, entering into states of ecstasy and trance.

Go to that impenetrable realm
That death himself trembles to look upon.
There plays the fountain of love
With swans sporting on its waters.

Conflict with Family

However her new family did not approve of her piety and devotion to Krishna. To make things worse Mira refused to worship their family deity Durga. She said she had already committed herself to Sri Krishna. Her family became increasingly disproving of her actions, but the fame and saintly reputation of Mirabai spread throughout the region. Often she would spend time discussing spiritual issues with Sadhus and people would join in the singing of her bhajans. However this just made her family even more jealous. Mira's sister-in-law Udabai started to spread false gossip and defamatory remarks about Mirabai. She said Mira was entertaining men in her room. Her husband, believing these stories to be true, tore into her room with sword in hand. However he saw Mira only playing with a doll. No man was there at all.

Mirabai and Akbar

Mira's fame spread far and wide her devotional bhajans were sung across northern India. It is said that the fame and spirituality of Mirabai reached the ears of the Moghul Emperor Akbar. Akbar was tremendously powerful, but he was also very interested in different religious paths. The problem was that he and Mirabai's family were the worst enemies; to visit Mirabai would cause problems for both him and Mirabai. But

Akbar was determined to see Mirabai, the Princess – Saint. Disguised in the clothes of beggars he travelled with Tansen to visit Mirabai. Akbar was so enamoured of her soulful music and devotional singing, that he placed at her feet a priceless necklace before leaving. However in the course of time Akbar's visit came to the ears of her husband Bhoj Raj. He was furious that a Muslim and his own arch enemy and set eyes upon his wife. He ordered Mirabai to commit suicide by drowning in a river. Mirabai intended to honour her husbands command, but as she was entering the river Sri Krishna appeared to her and commanded her to leave for Brindaban where she could worship him in peace. So with a few followers, Mirabai left for Brindaban, where she spent her time in devotion to Sri Krishna. After a while her husband became repentant, feeling that her wife was actually a real saint.

Poems of Mirabai

Much of what we know about Mirabai comes from her poetry. Her poetry express the longing and seeking of her soul for union with Sri Krishna. At time she expresses the pain of separation and at other times the ecstasy of divine union. Her devotional poems were designed to be sung as bhajans and many are still sung today. "Mira's songs infuse faith, courage, devotion and love of God in the minds of the readers. They inspire the aspirants to take to the path of devotion and they produce in them a marvelous thrill and a melting of the heart." Mirabai was a devotee of the highest order. She was immune to the criticism and suffering of the world. She was born a princess but forsook the pleasures of a palace for begging on the streets of Brindaban. She lived during a time of war and spiritual decline, but her life offered a shining example of the purest devotion.

It is said in her death she melted into the heart of Krishna. Tradition relates how one day she was singing in a temple, when Sri Krishna appeared in his subtle form. Sri Krishna was so pleased with his dearest devotee. He opened up his heart centre and Mirabai entered leaving her body whilst in the highest state of Krishna consciousness.

97

Lady Gaga

Stefani Joanne Angelina Germanotta (born March 28, 1986), better known by her stage name Lady Gaga, is an American pop singer. She began performing in the rock music scene of New York City's Lower East Side in 2003 and enrolled at New York University's Tisch School of the Arts. She soon signed with Streamline Records, an imprint of Interscope Records. During her early time at Interscope, she worked as a songwriter for fellow label artists and captured the attention of Akon, who recognized her vocal abilities, and signed her to his own label, Kon Live Distribution. Released on August 19, 2008, her debut album, The Fame, reached number one in the UK, Canada, Austria, Germany and Ireland as well as accomplishing positions within the top-ten in numerous countries worldwide; in the United States, it peaked at number two on the Billboard 200 chart and topped Billboard's Dance/ Electronic Albums chart. Its first two singles, "Just Dance" and "Poker Face", co-written and co-produced with RedOne, became international number-one hits, topping the Billboard Hot 100 in the United States as well as the charts of other countries. The album, which later earned a total of six Grammy Award nominations, won the awards for Best Electronic/Dance Album and Best Dance Recording for "Poker Face".

In early 2009, she embarked on her first headlining tour, The Fame Ball Tour. By the fourth quarter of the year, she had released her EP, The Fame Monster, along with the global chart-topping lead single "Bad Romance", as well as having embarked on her second headlining tour of the year, The Monster Ball Tour. Her second album, Born This Way, is scheduled for release in 2011. Lady Gaga is inspired by glam rock artists such as David Bowie and Queen, as well as pop singers such as Madonna and Michael Jackson. She has also stated fashion is a source of inspiration for her songwriting and performances. Gaga was ranked the 73rd Artist of the 2000–10 decade by Billboard. As of August 2010, Gaga has sold more than 15 million albums and 51 million singles worldwide. Both Time magazine and Forbes included Gaga in its annual Time 100 list of the most influential people in the world and the 100 Most Powerful and Influential celebrities in the world, respectively. Forbes also placed her at number seven on their annual list of the World's 100 Most Powerful Women. Gaga has been influenced by glam rock artists such as David Bowie and Queen, as well as pop music artists such as Madonna, Britney Spears and Michael Jackson. The Queen song "Radio Ga Ga" inspired her stage name, "Lady Gaga". She commented: "I adored Freddie Mercury and Queen had a hit called 'Radio Gaga'. That's why I love the name [...] Freddie was unique – one of the biggest personalities in the whole of pop music." In response to the comparisons between herself and Madonna, Gaga stated: "I don't want to sound presumptuous, but I've made it my goal to revolutionise pop music. The last revolution was launched by Madonna 25 years ago." Actress and singer Grace Jones was also cited as an inspiration, along with Blondie singer Debbie Harry.

98

Pratibha Patil

Pratibha Devisingh Patil is the current President of India, the 12th person and first woman to hold the office. She was sworn in as President of India on July 25, 2007, succeeding Dr. A.P.J. Abdul Kalam. Patil, a member of the Indian National Congress (INC), was nominated by the ruling United Progressive Alliance and Indian Left. She won the presidential election held on July 19, 2007 defeating her nearest rival Bhairon Singh Shekhawat by over 300,000 votes. Pratibha Shekhawat was born to Narayan Rao Patil in Nadgaon, Maharashtra. She studied at R.R. School at Jalgaon. She received her M.A. from Mooljee Jaitha (M.J.) College, Jalgaon (affiliated to North Maharashtra University, Jalgaon) and obtained a law degree from the Government Law College, Mumbai (affiliated to University of Bombay). She married educator Devisingh Ramsingh Shekhawat on July 7, 1965. One of her colleagues and close friends at the University of Bombay, Nishant Raina, introduced her to table tennis, a sport she excelled in, winning various inter-college tournaments. In 1962, Pratibha Patil was voted \"College Queen\" of M.J. College. The same year, she won an assembly election from Edlabad constituency on the Indian National Congress ticket. Patil represented Edlabad constituency in Jalgaon District, Maharashtra as a member of the Maharashtra Legislative Assembly (1962-1985), and was deputy chairwoman of the Rajya Sabha (1986-1988), Member of Parliament from Amravati in the Lok Sabha (1991-1996), and the 24th, and the first woman Governor of Rajasthan (2004-2007).

She began her political career in 1962 at the age of 27. Under the mentorship of senior Congress leaders like Dr

Abasaheb, Gopalrao Khedkar and ex-Chief Minister Yashwantrao Chavan, she became a deputy minister for education after re-election in 1967 (in the Vasantrao Naik ministry). In her next terms (1972-78) she was a full cabinet minister for the state. In successive congress governments, she handled the portfolios of tourism, social welfare and housing under several chief ministers, Vasantdada Patil, Babasaheb Bhosle, S. B. Chavan and Sharad Pawar. She was continually re-elected to the assembly, either from Jalgaon or the nearby Edlabad constituencies, until 1985, when she was elected to the Rajya Sabha as a Congress candidate. She has never lost an election that she has contested. In 1977, the Congress party split up after Indira Gandhi's defeat following the Indian Emergency (1975?1977). Many senior leaders of state Congress(I), including Pratibha's mentor Chavan and his protege Sharad Pawar, as well as much of the rank and file joined the Congress (Urs) party floated by Devraj Urs. However, Pratibha preferred to remain with Indira Gandhi, though it verged on inviting political ridicule.

In 1980, the Congress (I) swept back into power, and her name was considered a front-runner for the Chief Minister's post. However, the post went to Sanjay Gandhi's confidant A. R. Antulay, who was soon forced to resign on corruption charges. Subsequently, she became a minister again in the Vasantdada Patil ministry. Following differences between Patil and then Maharashtra Pradesh Congress Committee (MPCC) chief Prabha Rau, Rajiv Gandhi appointed her as MPCC chief (1988-90). She was elected to the Rajya Sabha, and served as its deputy chairperson from November 1986 to November 1988. Her term expired in April 1990. The following year, in the elections when Rajiv Gandhi was assassinated she won the election for the 10th Lok Sabha from Amravati constituency, her husband's city, where he had once been mayor, thus first time entering in lower house of national parliament Lok Sabha. She has also served as Director of National Federation of Urban Co-operative Banks & Credit Societies and the Member of Governing Council, National Co-operative, Union of India.

She was appointed as Governor of Rajasthan in November 2004, eight years after she had completed her term in the

10th Lok Sabha, Pratibha Patil was recalled from political hibernation to become the first woman Governor of Rajasthan. She was the second politician from Maharashtra in this post, the first being Vasantdada Patil. With Pratibha Patil as Governor, Rajasthan had women in three significant positions of power in the state, including Chief Minister Vasundhara Raje and Assembly Speaker Sumitra Singh. In April 2006, the Rajasthan Legislative Assembly passed the Rajasthan Freedom of Religion Bill 2006. The objective of the bill was to control" unlawful conversion from one religion to another by allurement or by fraudulent means or forcibly." However, some Christian organizations opposed the bill alleging that the bill was an outcome of the rightist policies of Sangh Parivar. The Government of Rajasthan re-sent the bill to her in May 2006 noting that similar anti-conversion laws enacted by Congress governments in Madhya Pradesh and Orissa over 40 years ago were upheld by the Supreme Court of India and that the head of the Constituent Assembly, Dr B R Ambedkar, while drafting Article 25 of the Constitution had said that it would be best to leave it to the state legislatures to make laws to regulate conversions.

A similar bill named Himachal Pradesh Freedom of Religion Act 2006 passed later was promptly signed by the state governor.She resigned as the Governor of Rajasthan on June 21, 2007, due to her presidential candidacy. On 14 June, United Progressive Alliance (UPA), the ruling alliance of political parties in India headed by Congress (I), and the Indian left nominated her as their candidate for the Presidential Election to be held on 19 July 2007.She emerged as a compromise candidate after the Left parties would not agree to the nomination of past Home Minister Shivraj Patil. At that point, Sonia Gandhi proposed Pratibha Patil's name. Her loyalty to Nehru-Gandhi family was widely perceived to be a major factor in her nomination as UPA-Left Presidential candidate.She won the presidential election held on July 19, 2007 defeating her nearest rival Bhairon Singh Shekhawat by over 300,000 votes. She took office as India's first female president on July 25, 2007.

99

Boudica

Boudica was a famous queen of ancient Briton who led a rebellion against the Roman occupiers. Boudica was born in 30 AD in South East England. Around CE 48, she married Prasutagus, the head of the Iceni tribe in south East England. They lived in Norfolk and during the life of Prasutagus were given semi independence from the Roman occupiers. Prasutagus was given the freedom to remain King of the Iceni, but under the dominion of Rome. Despite some advantages of Roman rule, the Iceni people suffered many indignities such as slavery and high taxes. On the death of Prasutagus, Roman law meant the majority of his possessions would pass to the Roman Emperor. However, the local Roman commanders took this as a pretext to confiscate all the property of Prasutagus and other leading Iceni tribe members. Prasutagus had also run up debts during his lifetime and when his wife Boudica could not meet them, she was stripped and beaten in public. The Roman historian, Tacitus, wrote that Roman soldiers raped her daughters. Other tribes such as the Trinobantes were subject to similar treatment, leading to growing feelings of rebellion amongst the native Britons. It was Boudica who was able to unite the various warring tribes of Briton and lead them in revolt against the Roman occupiers. The Roman writer Cassius Dio described Boudica as:

> *"very tall. Her eyes seemed to stab you. Her voice was harsh and loud. Her thick, reddish-brown hair flung down below her waist. She always wore a great golden torc around her neck and a flowing tartan cloak fastened with a brooch."*

The first target for Boudica and the Britons was the Roman city of Colchester. This city stood as an emblem for

Roman rule; it housed a temple to the Roman emperor Claudius. The city was lightly defended and the Britons had little trouble in raising the city. The Roman governor, Suetonius, was at the time fighting in Anglesy. When he heard the news he travelled to London, which was then a small but thriving financial centre. He considered defending London, but concerned over the growing number of rebellious Britons he leaved it only lightly armed. The Britons, thus were able to take London and later St Albans. Over 80,000 people were estimated to have been killed in the 3 cities. The Britons showed no mercy to those left behind. Boudica now led her growing army north to meet Suetonius' army. Along the way Boudica's army were able to successfully ambush a Roman column of soldiers. The 2 armies eventually met in open battle, possibly at somewhere along Watling Street. The Britons, heavily outnumbered the Roman forces, however, the Romans had superior tactics, training, discipline and weapons. The Romans chose a narrow location, where the Britons could not use their superior numbers.

The first wave of Britons was stopped with a wave of Roman Javelins. When the second wave came, the Romans held firm, behind their wall of shields, stabbing the Britons with the short sword. At the end of the Battle, only 400 Romans had fallen, but up to 200,000 Britons had been slaughtered. Conflicting reports suggest that Boudica, either took poison or died from her wounds. Boudica gained a tremendous amount of interest in the Victorian period. The husband of Queen Victoria, Prince Albert, commissioned the statue of Boudica which stands outside the Houses of Parliament in London. Boudica, actually became a symbol of the British Empire, which is ironically given her anti imperialist stance. The rebellion of Boudica, almost forced the Romans to leave England, as it was the Emperor Nero decided to replace Suetonius with a more neutral ruler Publius Petronius Turpilianus. Most of the sources for Boudica's time comes from Tacitus. His father-in-law, Agricola, was a military tribune under Suetonius Paulinus, which almost certainly gave Tacitus an eyewitness source for Boudica's revolt.

100

Paula Radcliffe

Paula Jane Radcliffe, MBE (born 17 December 1973) is an English long-distance runner. She is the current women's world record holder in the marathon. In 2002, Radcliffe was voted BBC Sports Personality of the Year and was awarded an MBE.

Early Life

Radcliffe was born on 17 December 1973 in Davenham near Northwich, Cheshire. Her family then moved to nearby Barnton where she attended Little Leigh Primary School. Despite suffering from asthma and anaemia she took up running at the age of seven, influenced by her father who was a keen amateur marathon runner and joined Frodsham Athletic Club. Her family later moved to Kingsley. When Radcliffe was aged 12, the family moved to Oakley, Bedfordshire and she became a member of Bedford & County Athletics Club. Her joining the club coincided with a talented and dedicated coach, Alex Stanton, building the women's and girls' sections in to one of the strongest in the country, in spite of Bedford's relatively small size. Radcliffe's father became club vice-chairman and her mother, a fun-runner, managed the women's cross-country team. Her first race at a national level came as a 12 year old in 1986 when she placed 299th out of around 600 in the girls' race of the English Schools Cross Country Championships. She finished fourth in the same race one year later. Radcliffe attended Sharnbrook Upper School and Community College. She went on to study French, German and economics at Loughborough University, gaining a first-class honours degree in modern European studies.

Running Career

Early Career

Radcliffe's early running success was in cross country events, including the 1992 World Junior title, beating Wang Junxia. She missed the 1994 season through injury, but came back with a succession of good results at 5,000 m, including fifth place in both the 1995 World Championships and 1996 Olympic Games. Although a silver-medalist in the 1999 World Championships in Athletics Radcliffe finished out of the medals at the 2000 Olympic Games and 2001 World Championships. She won back-to-back titles in the 2000 and 2001 IAAF World Half Marathon Championships, and winning a third title in 2003.

Further World Records

Radcliffe is the current world record holder for the women's marathon, which she set during the 2003 London Marathon in April, with a time of 2:15:25. This mark is currently one of the highest scoring performances ever. In terms of IAAF world ranking points, at 1307, it is higher in value than Florence Griffith-Joyner's 100 and 200 m records, Marita Koch's 400 m, and Michael Johnson's 400 m record. Radcliffe is also the current world record holder for the women's road 10k in a time of 30 minutes and 21 seconds, which she set on 23 February 2003 in the World's Best 10K in San Juan, Puerto Rico. Radcliffe won the 2004 New York City Marathon in a time of 2:23:10, beating Kenya's Susan Chepkemei. Of the seven marathons Radcliffe has run so far, she has won six and set a record in five. She has run four out of the five fastest times in history in the women's marathon.

2004 Athens Olympics

Radcliffe did not compete in the London Marathon in 2004, but was the favourite to win a gold medal in the marathon at the Olympic Games in Athens. However, she suffered an injury to her leg just two weeks prior to the event and had to use a high dose of anti-inflammatory drugs. This had an adverse effect on her stomach hindering food absorption. She ended up withdrawing from the race after 36 km. Five days later she started in the 10,000 metres but, still suffering from the effects of the marathon, retired with eight laps remaining. Radcliffe said "You go through bad stages in a marathon, but never as bad as that", "I've never before not been able to finish and I'm desperately trying to find a reason for what happened", "I just feel numb - this is something I worked so hard for." Regarded as Great Britain's best gold medal hope in athletics, her withdrawal made headlines in the UK, with editorial stances ranging from support to negativity, with some newspapers deriding Radcliffe for 'quitting', rather than going on to finish the race. Television pictures showed Radcliffe in a clearly distressed state after dropping out of the marathon, being comforted by two friends from her early running days.

Family and Autobiography

Radcliffe took a break through the 2006 season owing to injuries and in July announced that she was expecting her first child. Her comeback was further delayed in 2007 as a result of a stress fracture in her lower back. Radcliffe chose not to defend her world marathon crown in 2007, in order to undertake further rehabilitation, but insisted she wanted to compete in the next two Olympics. She made her return to competitive running on 30 September 2007, taking part in the BUPA Great North Run in the UK on Tyneside. This was her first race in almost two years. Radcliffe finished second behind US runner Kara Goucher. Radcliffe made her marathon return at the New York City Marathon on 4 November 2007 which she won with an official time of 2:23:09. Radcliffe released an autobiography in 2007, *Paula: My Story So Far.*

Other Achievements and Awards

- Awarded the BBC London Sports Awards 2003 for 'Sporting Moment of the Year'.
- Radcliffe has set numerous records, official and unofficial, on the track and the roads. As of November 2009, she holds the official world record for 10 km on roads. She has twice won the World Half-Marathon championships, twice the World Cross-Country championships (in 2001 and 2002), and in December 2003 became European Cross-Country champion for the second time, the only woman to have achieved this feat in the event's ten-year history.

- In 2004 Radcliffe joined with Jonathan Edwards on an *Olympic Special Who Wants to Be a Millionaire?*. The pair raised £64,000 for charity, half of that sum going to the British Olympic Association and a quarter of the sum going to Asthma UK.
- Nominated for the Sports Personality Award in 2007.
- Won the Laureus World Comeback of the Year award in early 2008 for her performances in 2007.

Personal Life

Radcliffe married her coach, Northern Irish former international 1,500 m runner Gary Lough. in April 2000 in Bedford. At age 33, she gave birth to her first child. Her daughter Isla was born on 17 January 2007 at 9:43 a.m. at the Princess Grace Hospital in Monaco after a 27-hour labour. Her second child, a son, Raphael, was born on 29 September 2010.

101

Louisa May Alcott

Louisa May Alcott (November 29, 1832 – March 6, 1888) was an American novelist. She is best known for the novel *Little Women* and its sequels *Little Men* and *Jo's Boys*. *Little Women* was set in the Alcott family home, Orchard House in Concord, Massachusetts, and published in 1868. This novel is loosely based on her childhood experiences with her three sisters.

Childhood and Early Work

Alcott was the daughter of noted transcendentalist and educator Amos Bronson Alcott and Abigail May Alcott. She shared a birthday with her father on November 29, 1832. In a letter to his brother-in-law, Samuel Joseph May, a noted abolitionist, her father wrote: "It is with great pleasure that I announce to you the birth of my second daughter...born about half-past 12 this morning, on my [33] birthday." Though of New England heritage, she was born in Germantown, which is currently part of Philadelphia, Pennsylvania. She was the second of four daughters: Anna Bronson Alcott was the eldest; Elizabeth Sewall Alcott and Abigail May Alcott were the two youngest. The family moved to Boston in 1834, After the family moved to Massachusetts, Alcott's father established an experimental school and joined the Transcendental Club with Ralph Waldo Emerson and Henry David Thoreau.

In 1840, after several setbacks with the school, the Alcott family moved to a cottage on 2 acres (8,100 m) of land, situated along the Sudbury River in Concord, Massachusetts. The Alcott family moved to the Utopian Fruitlands community for a brief interval in 1843-1844 and then, after its collapse, to rented rooms and finally to a house in Concord purchased with her mother's inheritance and financial help from Emerson. They moved into the home they named "Hillside" on April 1, 1845. Alcott's early education included lessons from the naturalist Henry David Thoreau. She received the majority of her schooling from her father. She received some instruction also from writers and educators such as Ralph Waldo Emerson, Nathaniel Hawthorne, and Margaret Fuller, who were all family friends. She later described these early years in a newspaper sketch entitled "Transcendental Wild Oats". The sketch was reprinted in the volume *Silver Pitchers* (1876), which relates the family's experiment in "plain living and high thinking" at Fruitlands. As an adult, Alcott was an abolitionist and a feminist. In 1847, the family housed a fugitive slave for one week. In 1848, Alcott read and admired the "Declaration of Sentiments" published by the Seneca Falls Convention on women's rights.

Literary Success and Later Life

Alcott's literary success arrived with the publication by the Roberts Brothers of the first part of *Little Women: or Meg, Jo, Beth and Amy*, (1868) a semi-autobiographical account of her childhood with her sisters in Concord, Massachusetts. Part two, or *Part Second*, also known as *Good Wives*, (1869) followed the March sisters into adulthood and their respective marriages. *Little Men* (1871) detailed Jo's life at the Plumfield School that she founded with her husband Professor Bhaer at the conclusion of Part Two of *Little Women*. *Jo's Boys* (1886) completed the "March Family Saga". In *Little Women*, Alcott based her heroine "Jo" on herself. But whereas Jo marries at the end of the story, Alcott remained single throughout her life. She explained her "spinsterhood" in an interview with Louise Chandler Moulton, "because I have fallen in love with so many pretty girls and never once the least bit with any man." However, Alcott's romance while in Europe with Ladislas Wisniewski, "Laddie", was detailed in her journals but then deleted by Alcott herself before her death. Alcott identified Laddie as the model for Laurie in *Little Women*, and there is strong evidence this was the significant emotional relationship of her life. When her younger sister May died in 1879, Alcott took in May's daughter, Louisa May Nieriker ("Lulu"), who was two years old. The baby had been named after her aunt, but was nicknamed Lulu, whereas Louisa May's nicknames were "Weed" and "Louy".

Selected Works

- *The Inheritance* (1849, unpublished until 1997)
- *Flower Fables* (1849)
- *Hospital Sketches* (1863)
- *The Rose Family: A Fairy Tale* (1864)
- *Moods* (1865, revised 1882)
- *Morning-Glories and Other Stories* (1867)
- *The Mysterious Key and What It Opened* (1867)
- *Little Women* or *Meg, Jo, Beth and Amy* (1868)
- *Three Proverb Stories* (includes "Kitty's Class Day", "Aunt Kipp" and "Psyche's Art") (1868)
- *A Strange Island*, (1868)
- *Part Second* of *Little Women*, also known as "Good Wives" (1869)
- *Perilous Play*, (1869)
- *An Old Fashioned Girl* (1870)
- *Will's Wonder Book* (1870)
- *Aunt Jo's Scrap-Bag* (1872–1882)
- *Little Men: Life at Plumfield with Jo's Boys* (1871)
- "Transcendental Wild Oats" (1873)
- *Work: A Story of Experience* (1873)

Index

G

H

I

J

K

L

M

N

O

P

■■■